Gjergji Kasneci

Searching and Ranking in Entity-Relationship Graphs

Gjergji Kasneci

Searching and Ranking in Entity-Relationship Graphs

From Web Pages to Knowledge

Südwestdeutscher Verlag für Hochschulschriften

Impressum/Imprint (nur für Deutschland/ only for Germany)
Bibliografische Information der Deutschen Nationalbibliothek: Die Deutsche Nationalbibliothek verzeichnet diese Publikation in der Deutschen Nationalbibliografie; detaillierte bibliografische Daten sind im Internet über http://dnb.d-nb.de abrufbar.

Alle in diesem Buch genannten Marken und Produktnamen unterliegen warenzeichen-, marken- oder patentrechtlichem Schutz bzw. sind Warenzeichen oder eingetragene Warenzeichen der jeweiligen Inhaber. Die Wiedergabe von Marken, Produktnamen, Gebrauchsnamen, Handelsnamen, Warenbezeichnungen u.s.w. in diesem Werk berechtigt auch ohne besondere Kennzeichnung nicht zu der Annahme, dass solche Namen im Sinne der Warenzeichen- und Markenschutzgesetzgebung als frei zu betrachten wären und daher von jedermann benutzt werden dürften.

Verlag: Südwestdeutscher Verlag für Hochschulschriften Aktiengesellschaft & Co. KG
Dudweiler Landstr. 99, 66123 Saarbrücken, Deutschland
Telefon +49 681 37 20 271-1, Telefax +49 681 37 20 271-0
Email: info@svh-verlag.de
Zugl.: Saarbruecken, Universitaet des Saarlandes, 2009

Herstellung in Deutschland:
Schaltungsdienst Lange o.H.G., Berlin
Books on Demand GmbH, Norderstedt
Reha GmbH, Saarbrücken
Amazon Distribution GmbH, Leipzig
ISBN: 978-3-8381-1761-4

Imprint (only for USA, GB)
Bibliographic information published by the Deutsche Nationalbibliothek: The Deutsche Nationalbibliothek lists this publication in the Deutsche Nationalbibliografie; detailed bibliographic data are available in the Internet at http://dnb.d-nb.de.

Any brand names and product names mentioned in this book are subject to trademark, brand or patent protection and are trademarks or registered trademarks of their respective holders. The use of brand names, product names, common names, trade names, product descriptions etc. even without a particular marking in this works is in no way to be construed to mean that such names may be regarded as unrestricted in respect of trademark and brand protection legislation and could thus be used by anyone.

Publisher: Südwestdeutscher Verlag für Hochschulschriften Aktiengesellschaft & Co. KG
Dudweiler Landstr. 99, 66123 Saarbrücken, Germany
Phone +49 681 37 20 271-1, Fax +49 681 37 20 271-0
Email: info@svh-verlag.de

Printed in the U.S.A.
Printed in the U.K. by (see last page)
ISBN: 978-3-8381-1761-4

Copyright © 2010 by the author and Südwestdeutscher Verlag für Hochschulschriften Aktiengesellschaft & Co. KG and licensors
All rights reserved. Saarbrücken 2010

Contents

1 Introduction   **1**
  1.1 Motivation 1
      1.1.1 Limits of Current Search Technology 1
      1.1.2 Opportunities 2
  1.2 Expressive Search with Entities and Relationships 3
  1.3 Challenges 5
  1.4 Contributions 6
  1.5 Outline 6

2 From Web Pages to Knowledge   **9**
  2.1 Harvesting Knowledge from the Web 9
      2.1.1 The Statistical Web 11
      2.1.2 The Semantic Web 11
      2.1.3 The Social Web 12
  2.2 Unifying The Social and the Semantic Web 13
      2.2.1 YAGO 14
  2.3 Summary 16

3 Entity-Relationship Graphs for Knowledge Representation   **17**
  3.1 Basics 17
  3.2 RDF/RDFS and OWL 18
  3.3 Storing and Querying Entity-Relationship Graphs 19
      3.3.1 Relational, Object-Oriented, and XML Databases 19
      3.3.2 Storing RDF data 22
      3.3.3 SPARQL 22
  3.4 Summary 23

4 NAGA   **25**
  4.1 Overview 25
      4.1.1 Motivation and Problem Statement 25
      4.1.2 Related Work 29
      4.1.3 Contributions and Outline 33
  4.2 A Framework for Querying with Entities and Relationships 33
      4.2.1 Query Model 33
      4.2.2 Answer Model 34

		4.2.3	Simple-Relationship Queries	35
		4.2.4	Regular-Expression Queries	36
	4.3	A Framework for Ranking with Entities and Relationships		37
		4.3.1	Ranking Desiderata	37
		4.3.2	Statistical Language Models for Document Retrieval	38
		4.3.3	The NAGA Ranking	40
		4.3.4	Related Ranking Models	44
	4.4	The NAGA Engine		45
		4.4.1	Architecture	45
		4.4.2	Implementation	46
		4.4.3	Efficiency Aspects	48
	4.5	Experimental Evaluation		49
		4.5.1	Setup	49
		4.5.2	Measurements	52
		4.5.3	Results and Discussion	53
	4.6	Conclusion		55
5	**STAR**			**57**
	5.1	Overview		57
		5.1.1	Motivation and Problem Statement	57
		5.1.2	Related Work	59
		5.1.3	Contributions and Outline	64
	5.2	The STAR Algorithm		65
		5.2.1	The STAR Metaheuristics	65
		5.2.2	First Phase: Quick Construction of Initial Tree	66
		5.2.3	Second Phase: Searching for Better Trees	67
		5.2.4	Discussion	73
	5.3	Approximation Guarantee		74
	5.4	Time Complexity		77
	5.5	Approximate Top-k Interconnections		78
	5.6	Experimental Evaluation		80
		5.6.1	Top-1 Comparison of STAR, DNH, DPBF, and BANKS	80
		5.6.2	Top-k comparison of STAR, BANKS, and BLINKS	82
		5.6.3	Comparison of STAR and BANKS	84
		5.6.4	Summary of Results	85
	5.7	Conclusion		86
6	**MING**			**87**
	6.1	Overview		87
		6.1.1	Motivation and Problem Statement	87
		6.1.2	Related Work	90
		6.1.3	Contributions and Outline	92

6.2	ER-Based Informativeness		92
	6.2.1	Statistics-Based Edge Weights	94
	6.2.2	IRank for Node-Based Informativeness	96
	6.2.3	Most Informative Subgraphs	97
6.3	The MING Algorithm		99
	6.3.1	First Phase: Candidate Subgraph Generation	100
	6.3.2	Second Phase: Mining the Most Informative ER Subgraph	101
6.4	Experimental Evaluation		105
	6.4.1	Efficiency	106
	6.4.2	Quality	108
6.5	Conclusion		111

7 Conclusion   **113**

8 Appendix   **115**
    8.1 Queries for the User Evaluation of NAGA 115
    8.2 MING Queries for the User Evaluation 117

Abstract

The Web bears the potential to become the world's most comprehensive knowledge base. Organizing information from the Web into entity-relationship graph structures could be a first step towards unleashing this potential. In a second step, the inherent semantics of such structures would have to be exploited by expressive search techniques that go beyond today's keyword search paradigm. In this realm, as a first contribution of this work, we present NAGA (**N**ot **A**nother **G**oogle **A**nswer), a new semantic search engine. NAGA provides an expressive, graph-based query language that enables queries with entities and relationships. The results are retrieved based on subgraph matching techniques and ranked by means of a statistical ranking model.

As a second contribution, we present STAR (**S**teiner **T**ree **A**pproximation in **R**elationship Graphs), an efficient technique for finding "close" relations (i.e., compact connections) between $k (\geq 2)$ entities of interest in large entity-relationship graphs.

Our third contribution is MING (**M**ining **In**formative **G**raphs). MING is an efficient method for retrieving "informative" subgraphs for $k (\geq 2)$ entities of interest from an entity-relationship graph. Intuitively, these would be subgraphs that can explain the relations between the k entities of interest. The knowledge discovery tasks supported by MING have a stronger semantic flavor than the ones supported by STAR.

STAR and MING are integrated into the query answering component of the NAGA engine. NAGA itself is a fully implemented prototype system and is part of the YAGO-NAGA project.

Summary

The World Wide Web bears the potential to become the world's most comprehensive knowledge base, but current keyword-based search technology is far from exploiting this potential. For example, suppose that we are interested in a comprehensive list of politicians who are also scientists. First, it is close to impossible to formulate our search need in terms of keywords. Second, the answer is possibly distributed across multiple pages, so that no state-of-the-art search engine will be able to find it. In fact, posing this query to Google (by using the keywords "scientist" and "politician") yields mostly news articles about science and politics. This example highlights the need for new, more expressive search techniques, as well as for explicit, unifying structures for the information on the Web.

There are various efforts that are aiming to add semantics to the Web by organizing information from the Web into entity-relationship-aware structures (e.g., YAGO [137, 138, 136], DBpedia [24], the Linking Open Data Project [30], Freebase [4], OpenCyc [56], etc.). The results of these efforts are large knowledge bases, organized as entity-relationship graphs, with explicit facts about entities (such as persons, organizations, locations, dates, etc.) and relationships (such as *isA*, *bornOnDate*, *locatedIn*, etc.). Motivated by these efforts, we address the problem of advanced knowledge search with entities and relationships. More specifically, we address the following problems:

1. Expressing and answering advanced knowledge queries about entities and relationships, e.g.: "Which physicists had Max Planck as academic advisor and what prizes have they won?"

2. Ranking the retrieved answers; an insightful ranking must prioritize answers about important entities.

3. Efficient discovery of "close" or "insightful" relations between $k(\geq 2)$ entities of interest. These kinds of knowledge discovery tasks aim at capturing the connections that can explain the relations between the k entities of interest.

In consideration of these problems, we propose **NAGA** [98, 97, 99] (Not Another Google Answer) as a first contribution of this work. NAGA is a new semantic search engine that is geared for large knowledge bases, which are organized as entity-relationship graphs. A graph-based query language enables the formulation of expressive queries with entities and relationships. The retrieved results are subgraphs from the knowledge base that match the query structure. They are ranked by a statistical ranking mechanism based on the principles of generative language models. For the returned answers, our ranking framework formalizes several intuitive desiderata such as *confidence*, *informativeness*, and *compactness*. The confidence reflects the correctness of results. The informativeness captures the importance of answers, and the compactness favors tightly connected entities in the answers. NAGA's superior result quality is demonstrated in comparison to state-of-the-art search engines and question answering systems.

Finding "close" relations between two, three, or more entities of interest is an important building block for many search, ranking, and analysis tasks. In large entity-relationship graphs with millions of nodes and edges, these kinds of tasks are computationally very challenging. In fact, from a graph-theoretic point of view, the underlying problem translates into the Steiner tree problem, which is known to be NP-hard. Intuitively, a Steiner tree that interconnects the given entities of interest represents the "closest" relations between them.

For this problem, we propose a new approximation algorithm coined **STAR** [95] (**S**teiner **T**ree **A**pproximation in **R**elationship Graphs). For n query entities, STAR yields an $O(log(n))$ approximation of the optimal Steiner tree in pseudopolynomial runtime. Furthermore, for practical cases, the results returned by STAR are qualitatively comparable to, or even better than, the results returned by a classical 2-approximation algorithm. STAR is extended to retrieve the approximate top-k Steiner trees for n given query entities. We have evaluated STAR over both main-memory as well as completely disk-resident graphs containing millions of nodes and tens of millions of edges. Our experiments show that in terms of efficiency STAR outperforms the best state-of-the-art database methods by a large margin, and also returns qualitatively better results.

A semantically more challenging knowledge discovery scenario is the one of finding a subgraph that can explain the relations between two or more entities of interest from a large entity-relationship graph. We refer to such subgraphs as *informative* subgraphs. This problem of finding informative subgraphs is more general than the one addressed by STAR, in that it considers whole subgraphs and not only trees. It is semantically more challenging than the problem addressed by STAR, in that we have to think of an adequate measure that favors insightful and salient relations between the query entities.

For this problem we propose **MING** [94, 93] (**M**ining **In**formative **G**raphs), an efficient method for finding and extracting an informative subgraph for $k (\geq 2)$ query entities. MING builds on a framework for computing a new notion of informativeness of nodes in entity-relationship graphs. This is used for computing the informativeness of entire subgraphs. The viability of our approach is demonstrated through experiments on real-life datasets, with comparisons to prior work.

STAR and MING are both integrated into the query answering component of the NAGA search engine. NAGA itself is a fully implemented prototype system and is part of the YAGO-NAGA project [17].

Acknowledgements

First and foremost, I would like to thank my family for their love, persistent support, motivation, guidance and inspiration throughout all my endeavors.

This work would not have been possible without the scientific advice and consistent motivation of my supervisor and mentor Prof. Dr.-Ing. Gerhard Weikum. I would like to thank him for the opportunities and the scientific guidance he gave me. Furthermore, I would like to thank my colleagues Georgiana Ifrim and Fabian Suchanek with whom I had many inspiring and fruitful scientific and philosophical discussions. Many other people with whom I have collaborated deserve my thanks, among them, Maya Ramanath, Mauro Sozio, and Shady Elbassuoni.

I owe many thanks to the International Max-Planck Research School (IMPRS) for my financial support, which allowed me to concentrate on my research.

Last, but certainly not least, I would like to thank the authors of [82] and the authors of [28, 92] for providing us with the Java code of their methods, BLINKS and BANKS, and the authors of [61] for providing us with the original C++ code of their method, DPBF.

1 Introduction

We are often faced with great opportunities brilliantly disguised as impossible situations.

CHARLES R. SWINDOLL

1.1 Motivation

1.1.1 Limits of Current Search Technology

Simple Boolean queries over title and abstract catalogs in libraries gave rise to a whole new field of Computer Science. This field is known today by the name of Information Retrieval. Since then, search technology has gone a long way. Today's search systems index billions of Web pages. They exploit information retrieval techniques on rich page features to satisfy the daily needs of hundreds of millions of users all around the globe.

The advances in search technology, however, concern mainly the retrieval of information in unstructured textual data, where the search paradigm is merely based on keywords. This search paradigm works well for keywords that need not be interpreted; but sometimes we are interested in explicit knowledge about entities and relationships holding between them. For example, consider the query that asks for prizes won by physicists who had Max Planck as academic advisor. No matter which keywords we use to express this query, current keyword search engines are not able to understand its intended meaning. For example, searching for the keywords "prize physicist Max Planck academic advisor" with Google yields mainly pages about Max-Planck Institutes or the Max Planck Society in the top-10 results. None of the top-10 results matches our query.

These kinds of queries pose several problems to keyword search engines. First, keywords cannot express advanced user needs that build on entities and relationships. Second, keyword search engines will do their best in trying to find Web pages that contain the query keywords. In our example, however, the result may be distributed across multiple pages, so that no state-of-the-art search engine will be able to find it. Third, not only the search but also the ranking strategies of current search engines are page-oriented. Searching with entities and relationships calls for new, more fine-grained ranking strategies that combine measures about the quality of pages with measures about the importance of entities and relationships in those pages.

1.1. Motivation

1.1.2 Opportunities

The above example highlights the need for more semantics and context awareness for the information organization and the search on the Web.

The quest for more semantics in the Web has attracted the attention of several research avenues of Computer Science such as Information Retrieval, Natural Language Processing, Information Extraction, Semantic and Social Web, Databases, etc. This research has ignited numerous projects with ambitious goals such as semantic annotation and editing of information [105, 3, 4], entity-centric information extraction and search [25, 35, 41, 120, 153], automatic construction and interlinking of general purpose knowledge bases [138, 24, 30, 56], community-based generation and combination of type-specific facts [59, 134, 148, 149], etc.

The semantic annotation of information and its organization in entity-relationship-aware structures opens up great opportunities for new entity-oriented search strategies. Some of these strategies are already being exploited in terms of faceted search, vertical-domain search, entity search, Deep-Web search, etc. All major search engines recognize a large fraction of product or company names, have built-in knowledge about geographic locations, and can return high-precision results for popular queries about consumer interests, traveling, and entertainment. Google, for example, understands entities based on the search context. When searching for "GM" Google returns pages about "General Motors", the query "GM food", on the other hand, yields pages about genetically modified food. Information-extraction and entity-search methods are clearly at work here. But these efforts seem to be focusing on specific domains only and do not exploit the notion of relationships.

Projects such as True Knowledge [12], Yahoo! Answers [150], Wolfram Alpha [15], Powerset [9] or START [10, 79] see in this realm a greater opportunity. They exploit Natural Language Processing in combination with background knowledge to answer natural language questions. However, the techniques behind these projects are not yet mature. All mentioned question answering engines have often problems understanding or dealing with questions for which the answer has to be composed from different pieces of information distributed across multiple pages. For example, none of these question answering systems can answer the question about prizes won by physicists who had Max Planck as academic advisor.

The opportunities that have guided this work and especially the YAGO-NAGA project [96, 17] are the following:

- We see the possibility of casting valuable parts of the Web information (i.e., information about science, culture, geography, etc.) into a consistent knowledge base that is organized as an entity-relationship graph. The nodes of such a graph would represent entities and the edges would stand for relationships holding between entities.

- Such an organization of information enables expressive and precise querying about entities and relationships. This can be exploited to make search more semantic, more knowledge-oriented, and less dependent on keywords or Web pages.

- We can take advantage of the redundancy of information in the Web to learn more about the importance of entities and relationships. This can be exploited to design new, more fine-grained ranking models that combine measures about the quality of Web pages with measures about the importance of entities and relationships in those pages.

- An entity-relationship-based organization of information from the Web together with a better understanding of importance at entity and relationship level paves the way for new, powerful analysis and knowledge discovery techniques.

1.2 Expressive Search with Entities and Relationships

The imprecise nature of queries in Information Retrieval makes us often feel uncomfortable, especially when our information needs are too intricate to be expressed through keywords. The gap between the user's information need and the query expressed through keywords is well-known. In contrast to the "uninterpreted" keyword search, the database community has given preference to precise query semantics. Query languages like SQL, for relational data, XQuery [49], for XML data, or SPARQL [54], for RDF graphs, have been proposed to deal with rigorous semantics. On the other hand, these query languages have little appeal for the end user. Hence, we believe that the next wave of search technology has to aim at understanding and answering natural language questions.

This work has mainly been driven by the vision of a search system that allows users to express their needs through queries that are formal counterparts of natural language questions. The basic elements of such queries are entities and relationships. The query language we have in mind is tailored for knowledge bases that are organized as entity-relationship graphs. It is akin to SPARQL, but it goes beyond SPARQL by supporting connectivity queries that ask for broad connections between entities or queries that capture the transitivity of relations such as *isA*, *partOf*, *locatedIn*, etc.

For example, consider the query that asks for philosophers from Germany who have been influenced by the English philosopher William of Ockham. We envision a query syntax that would allow us to formulate this query with entities and regular expressions over relationships. We give an example in the following.

William_of_Ockham influences x
x (bornIn|livesIn|isCitizenOf)locatedIn* Germany
x isa philosopher*

1.2. Expressive Search with Entities and Relationships

Without going into details, the term $x in the above query represents a variable that has to be bound with appropriate entities (i.e., philosophers from Germany who have been influenced by William of Ockham). The query, uses the regular expressions over relationships to express our search need without overspecifying it. For example, one can be generous when specifying that someone is from Germany by using the regular expression (*bornIn|livesIn|isCitizenOf*); the expression *locatedIn** helps capturing geographical hierarchies, e.g., with cities, counties, states, and countries. Similarly, the expression *influences** reflects that a philosopher may be directly or indirectly influenced by the philosophy of William of Ockham.

Such a query language would support the formulation of advanced search needs such as the ones reflected in the following examples.

- *Find a German Nobel Prize winner who survived both world wars and outlived all of his four children.*

  The answer is Max Planck. This search task illustrates the need for combining knowledge that may be distributed across multiple pages. The bits and pieces for the answer are not that difficult to locate: lists of Nobel prize winners, birth and death dates of these people, facts about family members extracted from biographies, etc. Gathering and connecting these facts is straightforward for a human, but it may take days of manually inspecting Web pages.

- *Find a comprehensive list of politicians who are also accomplished scientists.*

  Today's search engines fail on this kind of tasks, because they build on keyword matching techniques and cannot deal with entities, entity properties or relationships between entities. Additionally, the question entails a difficult ranking problem. Wikipedia alone contains hundreds of persons that are listed in the categories Politicians as well as Scientists. An insightful answer must rank important people first, for example, the German chancellor Angela Merkel who has a doctoral degree in physical chemistry, or Benjamin Franklin, and the like.

- *Find close relations between Renée Zellweger, Albert Einstein and Steve Ballmer.*

  An interesting and somewhat close relation is that all three of them are Swiss citizens. Albert Einstein studied in Switzerland and acquired the Swiss citizenship in the same year he gained his diploma, Renée Zellweger is of Swiss origin, and Steve Ballmer received an honorary Swiss citizenship a few years ago. This case again illustrates the need for combining facts from different Web sources. It also entails a ranking problem since long or trivial connections (e.g., that all three query entities are persons) may be rather non-satisfactory from a user's viewpoint.

The answers to these search tasks are not pages; rather, they are composed of explicit knowledge fragments, eventually extracted from different Web pages.

We refer to corresponding queries as "knowledge queries". A search system for knowledge queries has to reward the additional semantic information (given by the entities and the relationships) of the query by returning precise and salient answers. Whenever a query yields multiple answers, the system has to rank the most important answers first.

1.3 Challenges

The above search tasks highlight the need for more explicit, unifying structures for the information on the Web. Knowledge bases that organize information extracted from the Web as entity-relationship graphs are an important building block; but they are useless without a query language that exploits their inherent semantics.

The main challenges that have been addressed in this work are:

Expressive Querying: Designing an expressive query language that is tailored to information organized in entity-relationship graphs and allows the formulation of knowledge queries with entities and relationships. A prominent approach that addresses this challenge is SPARQL [54]. However, SPARQL does allow us to capture the transitivity of relations or broad connections between entities.

Ranking: Knowledge queries may often yield plenty of results. Hence the results need to be ranked. For example, the query that asks for a comprehensive list of German physicists may return hundreds of results. An insightful ranking has to give preference to important German physicists such as Albert Einstein, Max Planck, and the like. Ranking models for knowledge queries are much more difficult than traditional ranking models known from Information Retrieval, as one needs to reason about importance at entity and relationship level, and consider the semantics and the structure of both queries and results.

Efficient Search: Evaluating knowledge queries over graphs is computationally hard. Moreover, the need for ranking calls for smart evaluation strategies.

Efficient Knowledge Discovery: Especially challenging, from an efficiency and a semantics point of view, are queries that ask for commonalities or broad connections between two or more entities of interest. An example is the query that asks for the relations between Renée Zellweger, Albert Einstein and Steve Ballmer. These queries aim at knowledge discovery. From a semantics standpoint, one has to reason about measures that favor important connections between the entities of interest. From an efficiency standpoint, one has to think about algorithms that can efficiently discover these connections.

1.4 Contributions

This work contributes to advanced forms of search on entity-relationship graphs. We investigate a spectrum of issues ranging from expressive means for querying with entities and relationships to efficient knowledge discovery in entity-relationship graphs. Our main contributions are the following:

1. NAGA (**N**ot **A**nother **G**oogle **A**nswer).
   NAGA is a new semantic search engine. It provides an expressive, graph-based query language that supports queries about entities and relationships. The results are retrieved based on subgraph matching techniques and ranked by means of a statistical ranking model.

2. STAR (**S**teiner **T**ree **A**pproximation in **R**elationship Graphs).
   STAR is an efficient technique for finding "close" relations (i.e., compact connections) between $k(\geq 2)$ entities of interest in entity-relationship graph structures.

3. MING (**M**ining **In**formative **G**raphs).
   MING is an efficient method for retrieving "informative" subgraphs for $k(\geq 2)$ given entities of interest. Intuitively, these would be a subgraphs that can explain the relations between the entities of interest. In comparison to STAR, the knowledge discovery tasks supported by MING have a stronger semantic flavor. An adequate measure for informativeness should favor insightful and salient relations between the entities of interest (not necessarily compact ones).

The contributions presented in this work have been published or accepted for publication in various international conference proceedings and journals. The following paragraph gives an overview of the main publications.

Our work on NAGA has been published in the proceedings of WWW 2007 [99] and ICDE 2008 [98] and has been presented as a demo at SIGMOD 2008 [97]. An overview of the YAGO-NAGA project has been given in the December edition of SIGMOD Record 2008 [96]. The STAR algorithm has been published in the proceedings of ICDE 2009 [95], and the work on MING has been accepted for publication in the proceedings of CIKM 2009 [93].

1.5 Outline

The remainder of this work is organized as follows. In Chapter 2, we give a brief overview on research efforts towards extracting information from the Web and organizing it in high-quality knowledge bases. Along these lines, we present our own project YAGO (**Y**et **A**nother **G**reat **O**ntology). YAGO is a successful example for building high-quality knowledge bases that organize information from the Web

1.5. Outline

in entity-relationship graphs. Chapter 3 is dedicated to entity-relationship graphs and gives an overview of approaches for storing and querying them. In Chapter 4, we present NAGA, our semantic search system. Chapters 5 and 6 are about efficient knowledge discovery methods in large entity-relationship graphs. Chapter 5 introduces STAR, our algorithm for finding compact connections between $k(\geq 2)$ entities of interest, and Chapter 6 presents MING, our method for finding subgraphs that can explain the connections between $k(\geq 2)$ entities of interest. We conclude in Chapter 7.

2 From Web Pages to Knowledge

"We are drowning in information but starved for knowledge."

JOHN NAISBITT

2.1 Harvesting Knowledge from the Web

As the Web evolves, there are more and more Web sources in the spirit of Web 2.0, which allow users to semantically annotate information in a collaborative way. The annotations range from simple keywords or tags to detailed descriptions or articles. As these kinds of social tagging/editing communities are flourishing, the (slightly older) Semantic Web research avenue is aiming to give more structure to the Web information. For more than a decade, this research avenue has been pursuing various projects with the goal to build comprehensive Semantic-Web-style knowledge sources which structure information in terms of entities and relationships. Together with other Computer Science avenues such as Information Extraction and Databases, the Web 2.0 and the Semantic Web research avenue are contributing to the endeavor of adding more structure, more semantics, and more context-awareness to the information on the Web.

In this Chapter, we will explain why the efforts of these Web research avenues open up the great opportunity of "casting the Web into knowledge". The concrete idea is to extract high-quality information (in terms of data records) form the Web and store it into a consistent knowledge base. Such a knowledge base would contain explicit facts[1] about entities such as persons, locations, movies, dates, etc. The facts could be represented as relational tuples, RDF triples, or maybe XML fragments. Imagine a "Structured Web" that has the same scale and richness as the current Web but offers a precise and concise representation of knowledge stored in a knowledge base. This kind of Web would enable expressive and highly precise querying. Figure 1 illustrates a possible sample from such a knowledge base. While the nodes in the graph of Figure 1 represent entities the edges between them represent facts. Each fact may have a weight, intuitively representing the strength of the corresponding relationship between the two entities.

A knowledge base that contains the valuable information from the Web in a well structured form as above would support difficult queries that go beyond the capabilities of today's keyword-based search engines. Consider the HIV-relevant

---

[1] One can think of a fact as a structured data record.

2.1. Harvesting Knowledge from the Web

query that asks for a comprehensive list of drugs that inhibit proteases. Finding relevant answers to this query is extremely laborious and time-consuming, since one would have to browse through plenty of promising but eventually useless result pages. In order to increase the chance of retrieving better results, one could pursue the strategy of rephrasing the query; but this requires deep scholarly knowledge about the subject.

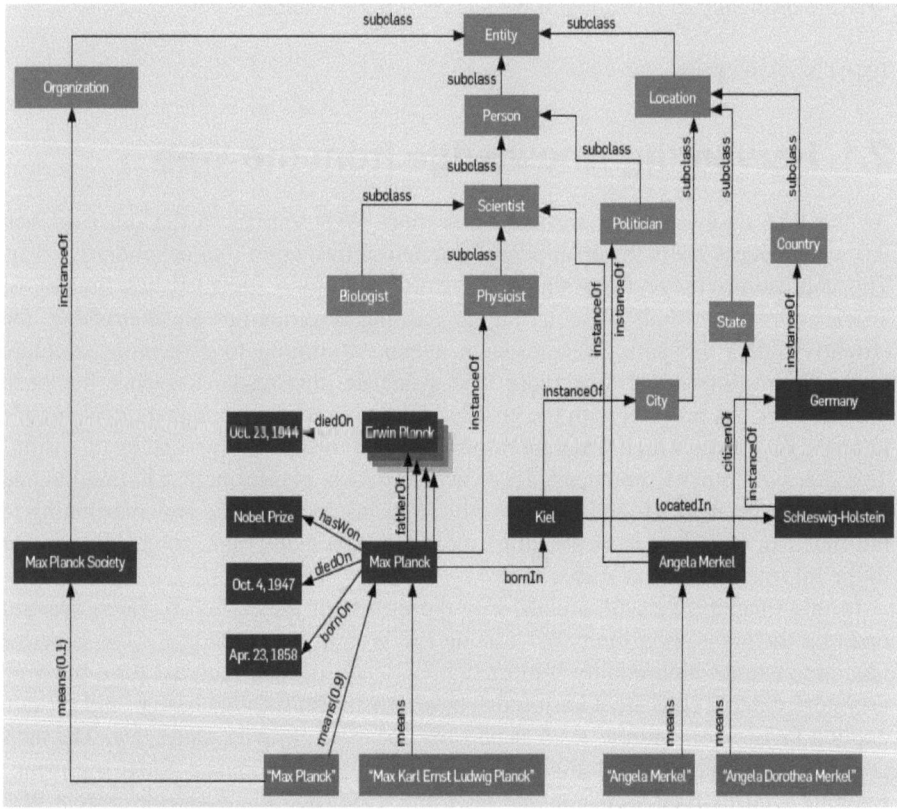

Figure 1: Sample knowledge graph (source [145])

Going one step further, such a knowledge base would also enable queries that ask for broad relations between entities. Consider the query that asks for commonalities or other connections between David Bohm, Niels Bohr, Enrico Fermi, and Richard Feynman. Close and interesting connections are that all four of them are quantum physicists, and that all four of them were members of the Manhattan Project. Discovering interesting relations between multiple entities on the Web is virtually

2.1. Harvesting Knowledge from the Web

impossible. An online answer generation would involve various difficult tasks such as analyzing huge amounts of Web pages, disambiguating entities, extracting and interconnecting facts, etc. Although the original task becomes simpler in a well structured knowledge base, it still remains difficult from an algorithmic point of view (see Chapter 5).

Other search tasks similar to the ones above were presented in Section 1.2.

There are three major Web research avenues which can contribute to the goal of casting the Web into explicit and clean knowledge: the *Statistical Web*, the *Semantic Web*, and the *Social Web*.

2.1.1 The Statistical Web

As of now, the greatest part of Web information still is in natural language text sources. The valuable scientific and cultural content of such sources is usually mixed up with ambiguous and noisy, low-quality information. Hence, the main challenge is to automatically extract clean, accurate and important facts from these kinds of sources. For such a large-scale information extraction task, one has to combine different natural language processing techniques (e.g., parsing, pattern matching, etc.) with statistical learning techniques (e.g., entity labeling, relational pattern learning, etc.) All these techniques have recently become more scalable and less dependent on human supervision [18, 55, 130]. However, extraction scalability and extraction accuracy are still two issues which seem to oppose each other. For example, the recent paradigm of *Machine Reading* [67], where the goal is to aggressively extract all possible binary relation instances from Web pages, helps to operate at a considerably larger scale, but often leads to inaccurate extraction. On the other hand, the *Life-long Learning* [25] paradigm, where the goal is to learn from already extracted information, helps to increase extraction accuracy but punishes the extraction efficiency. Hence, despite the continuous improvement, most of the extraction techniques still need to deal with efficiency and accuracy issues. Consequently, automatic information extraction techniques are not yet appropriate for the goal of extracting clean and accurate facts at Web scale.

2.1.2 The Semantic Web

Semantic-Web-style knowledge repositories like ontologies and taxonomies suggest a promising direction towards a machine processable Web. General-purpose ontologies and thesauri such as SUMO [121], OpenCyc [56], or WordNet [16, 72], provide comprehensive hierarchies of general concepts or classes such as computer scientist, physicist, musician, actor, etc. These hierarchies are usually built based on taxonomic relationships between entities, such as hyponymy and hypernymy (to describe relations between classes and subclasses), meronymy and holonymy (to describe relations between parts and a whole). Furthermore, these ontologies provide simple reasoning rules; for example they may know that humans are either

2.1. Harvesting Knowledge from the Web

male or female, cannot fly (without appropriate gadgets) but can compose and play music, and so on. Other domain-specifc ontologies and terminological taxonomies such as GeneOntology [5] or UMLS [14], in the biomedical domain, know specific domain entities (e.g., proteins, genes, etc.) and relations (e.g., specific biochemical interactions).

These kinds of knowledge sources have the advantage of satisfying the highest quality expectations, because they are manually assembled and curated. However, they are costly to build and continuous human effort is needed to keep them up to date. This negatively affects the coverage of such ontologies. For example WordNet is extremely rich in terms of classes but knows only few named entities (i.e., individuals) for each class. Furthermore, no hand-crafted ontology knows the most recent Windows version or the latest soccer star.

2.1.3 The Social Web

Social tagging and Web 2.0 communities, such as Wikipedia, flickr [3], Freebase [4], etc., which allow users to collaboratively annotate and edit information, constitute the Social Web. Human contributions are abundant in the form of semantically annotated Web pages, phrases in pages, images, or videos, together providing the "wisdom of the crowds". Recent endeavors along these lines are Freebase and Semantic Wikipedia. Inspired by Wikipedia, the Freebase project aims to construct an ontology by inviting volunteers to contribute by providing structured data records about entities or events. The Semantic Wikipedia project [105] is a comparable initiative. It invites Wikipedia authors to add semantic tags to their articles in order to turn the page link structure of Wikipedia into a large semantic network.

Apart from the large number of collaboratively added named entities (i.e., individuals) and annotations about them, Social Web sources can also provide high-quality information. In 2007, a study initiated on behalf of the German magazine "Stern", showed that due to more detailed and up-to-date information Wikipedia's quality was higher than the quality of the well-known German, universal lexicon "Brockhaus" [135]. Furthermore, a considerable part of this high-quality information is provided in semi-structured formats (e.g., Wikipadia infoboxes, lists, categories, etc.), which can be leveraged to extract high-quality facts about individuals.

Hence, both, the Semantic Web and the Social Web offer high-quality knowledge, and while the former has a large coverage on concepts (or classes), the latter has a large coverage on named entities (or individuals). For our goal of a clean and accurate knowledge base derived from the Web, the unification of these two sources seems to be very compelling.

In the next section, we give an example of how the knowledge provided by the Semantic and the Social Web can be combined into a large knowledge base.

2.2 Unifying The Social and the Semantic Web

There are various research projects which aim to combine elements from the three Web avenues mentioned above. The goals of these projects range from entity-centric fact extraction and search [25, 35, 41, 120, 153] to community-based generation and combination of type specific facts [59, 134, 148, 149]. Despite the great visions pursued by all these projects, in this section, we will shift our focus towards a more moderate endeavor. *Is it possible to turn relevant parts of the Social and the Semantic Web into a large knowledge base?*

In this section, we will present YAGO (**Y**et **A**nother **G**reat **O**ntology) [137, 138, 136] as a successful example of combining knowledge extracted from the Social Web with knowledge from the Semantic Web. YAGO is the first approach that successfully combines the goal of large-scale knowledge harvesting with the goal of maintaining a high accuracy and consistency.

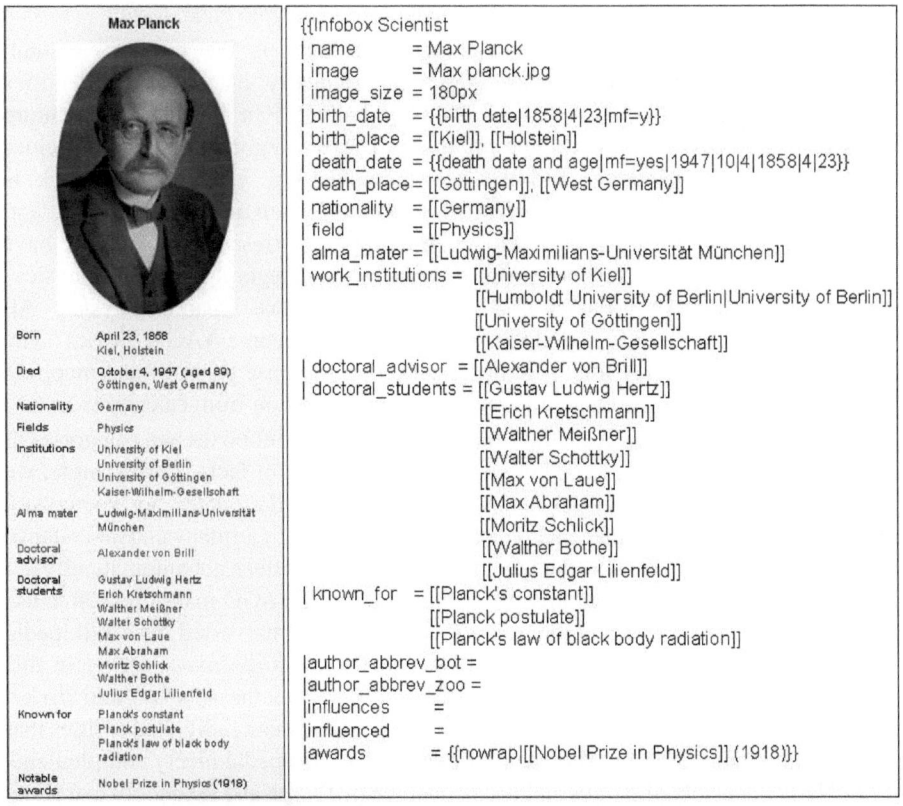

Figure 2: **The Wikipedia infobox of Max Planck**

2.2. Unifying The Social and the Semantic Web

2.2.1 YAGO

YAGO primarily gathers its knowledge by integrating information from Wikipedia and WordNet. Wikipedia provides various assets that can be seen as almost structured data: the infoboxes, the lists and the category system are such examples.

As depicted in Figure 2, infoboxes contain attribute names along with their values. The left hand side of the figure depicts the infobox as it appears on the Wikipedia page about Max-Planck. The editing format of the same infobox is depicted on the right hand side.

The editing format for infoboxes is based on templates which can be reused for important types of entities such as countries, companies, scientists, music bands, sports teams, etc. For example, the infobox of Max Planck gives us well-structured data about Max Planck's birth date, birth place, death date, nationality, alma mater, prizes, etc. It is almost straightforward to turn such an infobox into explicit facts. Consequently, we can extract clean facts about Max Planck, such as (*Max Planck, bornOn, 1858-04-23*) or (*Max Planck, hasWon, Nobel Prize*). YAGO provides automatic techniques for turning all Wikipedia infoboxes into clean facts.

Another Wikipedia asset for extracting clean facts is the category system. The Wikipedia article about Max Planck is manually placed into categories such as: "German Nobel laureates", "Nobel laureates in Physics", "Quantum physics", or "University of Munich alumni". These categories give YAGO clues about *instanceOf* relations, and we can infer that the entity Max Planck is an instance of the classes *GermanNobelLaureates*, *NobelLaureatesInPhysics*, and *UniversityOfMunichAlumni*. But when extracting the corresponding facts we have to be careful, as the placement of Max Planck in the category "Quantum physics" does not mean that Max Planck is an instance of the class *QuantumPhysics*. All Wikipedia categories extracted along with these facts become YAGO classes. The YAGO extractors employ linguistic processing (noun phrase parsing) and mapping rules, to achieve high accuracy in harvesting the information from categories.

The above examples indicate that solely relying on the infoboxes and categories of Wikipedia may result in a large but incoherent collection of facts. For example, we may know that Max Planck is an instance of *GermanNobelLaureates*, but we may not be able to automatically infer that he is also an instance of *Germans* and an instance of *NobelLaureates*. Likewise, the fact that he is a physicist does not automatically tell us that he is a scientist. To address these shortcomings, YAGO makes intensive use of the WordNet thesaurus and integrates the facts that are harvested from Wikipedia with the taxonomic backbone provided by WordNet. As discussed earlier in this chapter, WordNet knows many abstract classes as well as the *subclass* and *partOf* relations among them, but it has only sparse information about individual entities that would populate its classes. The wealth of entities in Wikipedia nicely complements WordNet; conversely, the rigor and high coverage of WordNet's taxonomy can make up for the gaps and noise in the Wikipedia category system. Hence, all WordNet

2.2. Unifying The Social and the Semantic Web

concepts become YAGO classes. More precisely, the whole *class-subclass* hierarchy of WordNet concepts is adopted into YAGO. Furthermore, all Wikipedia categories that become YAGO classes are mapped to the corresponding YAGO classes that were derived from WordNet. For example, the class *GermanNobelLaureates* is mapped to superclasses such as *NobelLaureates* or *Germans*.

YAGO also exploits Wikipedia's redirect system to infer words that refer to named entities. A Wikipedia redirect is a virtual Wikipedia page, which links to a real Wikipedia page. These links serve to redirect users to the correct Wikipedia article. For example, when the user types "Planck" or "Karl Ernst Ludwig Marx Planck" instead of "Max Planck", a virtual redirect page redirects the user to the Wikipedia page about Max Planck. From Wikipedia's redirect system YAGO extracts facts such as (*"Karl Ernst Ludwig Marx Planck"*, means, *Max Planck*). In fact, Figure 1 (Section 2.1) depicts an excerpt from YAGO. The edges between nodes highlighted in red or orange stand for facts about individuals; they were extracted from Wikipedia. The nodes highlighted in green correspond to YAGO classes. The edges between these nodes represent the *class-subclass* hierarchy. Edges between red and green nodes represent the *instanceOf* relation between individuals and YAGO classes.

QUALITY CONTROL YAGO pays particular attention to the consistency of the knowledge base.

When different extraction algorithms deliver the same fact, YAGO's consistency checking mechanism takes care to eliminate one of them. In case that one fact is more precise than another, then only the more precise fact is kept. For example, if the category harvesting has determined the birth year 1858 for Max Planck, and the infobox harvesting has determined 1858-04-23 as the date of birth, then only the more detailed fact with the date 1858-04-23 is kept. Furthermore, the consistency checking mechanism guarantees that the *subclass* relation remains acyclic.

YAGO relations are typed. For example the relationship *fatherOf* has the domain *Person* and the range *Person*. Hence, the fact *fatherOf(Max Planck, Quantum Theory)* would be discarded by YAGO's type-checking mechanism, since *Quantum Theory* is not a person. Furthermore, when a candidate fact contains an entity for which the extraction algorithm could not determine its class, it is discarded. Note that for all remaining facts, YAGO knows the class(es) and all superclasses for each entity.

Type checking can also be used to generate facts. For example, whenever a fact contains an unknown entity and the range or domain of the relation predicts that the entity should be a person, the algorithm makes the entity an instance of the class *Person*. In this case, a regular expression check is used to make sure that the entity name follows the basic pattern of given name and family name. The fact is kept only if the check was successful.

These and other procedures ensure that duplicate facts and dangling entities are removed, and the knowledge base remains consistent. YAGO is one of the largest knowledge bases available today; it contains around 2 million entities and about 20 million facts about them, where facts are instances of binary relations. Furthermore, its quality is unprecedented in the field of automatically generated ontologies. Extensive sampling showed that the accuracy is at least 95 percent, and many of the remaining errors (false positives) are due to incorrect entries in Wikipedia itself. Since its public release, YAGO has been adopted into several projects. For example, DBpedia [24], another project that aims to extract ontological data from Wikipedia, uses YAGO as a taxonomic backbone. YAGO has also been included into Freebase and is an integral part of the Linking Open Data Project [30], which aims to interconnect existing ontologies as Web services.

YAGO is publicly available at: `http://www.mpi-inf.mpg.de/yago-naga/yago/`.

2.3 Summary

In this chapter, we gave an overview of the evolutionary shift towards a Web with more structure, semantic flavor, and context-awareness. In this vision, we skimmed through various research domains that are taking the opportunities given by the Statistical, the Semantic and the Social Web, aiming to turn the Web into explicit knowledge. We saw that considerable progress in these research domains is often accompanied by limitations which mainly concern the quality and the scalability of information extraction, especially in the domain of the Statistical Web. Finally, by means of the YAGO project we demonstrated the viability of the idea of reconciling the Semantic and the Social Web into a high-quality knowledge base.

YAGO-style knowledge bases give rise to more expressive and precise querying, releasing the user from the restricted paradigm of keyword search, and paving the way towards question answering. The queries we have in mind would be formal counterparts of natural language questions. They would be entity-relation-based and would go beyond Datalog [37] or SPARQL-based [54] queries. But before diving into this topic, we will take a closer look at entity relationship graphs for knowledge representation.

3 Entity-Relationship Graphs for Knowledge Representation

"Perfection is reached, not when there is no longer anything to add, but when there is no longer anything to take away."

ANTOINE DE SAINT-EXUPÉRY

3.1 Basics

An important step towards representing knowledge from a domain of discourse is that of structuring the available information. For machine processable knowledge representation, the aim is to determine the resources associated with the explicit elements of the domain, such as entities and relationships. While a type-level representation aims at modeling classes of entities and their relationships, an instance-level representation aims at modeling the named entities (i.e., individuals) and their relationships. Often, the type-level and the instance-level modeling are combined. For example, in the previous chapter, we saw how YAGO combines the type-level representation of WordNet classes with the instance-level representation of Wikipedia entities.

Once the elements of discourse are determined, an entity-relationship graph can be built.

DEFINITION 1: [*Entity-Relationship Graph*]
Let Ent and Rel be finite sets of entity and relationship labels respectively. An entity-relationship graph *over* Ent *and* Rel *is a multigraph* $G = (V, l_{Ent}, E_{Rel})$ *where* V *is a finite set of nodes,* $l_{Ent} : V \to Ent$ *is an injective function, and* $E_{Rel} \subseteq l_{Ent}(V) \times Rel \times l_{Ent}(V)$ *is a set of labeled edges.*

While the labeled nodes of an entity-relationship graph (ER graph for short) correspond to entities, the lebeled edges represent relationships between entities. A sample ER graph is depicted in Figure 1 (Chapter 2).

A specific variant of type-level ER graphs for representing the conceptual schema of relational databases was introduced 1976 by Peter Chen [40]. This kind of graphs have been the undisputed light-weight model for semantic data representation and have significantly influenced various fields of computer science, such as software engineering, database modeling, artificial intelligence, and many more.

There are three main reasons for the ubiquity of ER graphs:

1. They are explicit (by means of node and edge labels),
2. They are simple, i.e., they structure information similarly to the way we do it in our minds; unnecessary information is omitted,
3. They are flexible, i.e., when representing schema-free information, edges can be easily added or removed.

Modern applications use ER graphs to represent knowledge from large domains such as Web-based social networks, biochemical networks, networks of products and customers, etc. Often these graphs are too large to fit in main memory. Hence, they need to be stored and manipulated on disk. Before discussing possibilities for storing and managing large ER graphs, we will present two prominent languages for ER-based knowledge representation.

3.2 RDF/RDFS and OWL

The vision of Semantic Web, with common formats for automatic data integration and manipulation, gave rise to two prominent ER-based languages for knowledge representation: RDFS and OWL.

RDF/RDFS The Resource Description Framework Schema (RDFS) is an extensible knowledge representation language recommended by the World Wide Web Consortium (W3C) [47] for the description of a domain of discourse (such as the Web). Its syntax is based on XML [46], and similarly to XML, RDFS allows the specification of a common syntax for data exchange. It enables the definition of domain resources, such as individuals, classes and properties. The basis of RDFS is RDF which comes with three basic symbols: *URIs* (Uniform Resource Identifiers) for uniquely addressing resources, *literals* for representing values such as string, numbers, etc., and *blank nodes* for representing unknown or unimportant resources. Another important RDF construct for expressing that two entities stand in a binary relationship is a *statement*. A statement is a triple of URIs and has the form (*Subject, Predicate, Object*), for example (*MaxPlanck, hasWonPrize, NobelPrize*). An RDF statement can be thought of as a an edge of an ER graph, where the *Subject* and the *Object* represent entity nodes and the *Predicate* represents the relationship label of the corresponding edge. In fact, the set of RDF statements about a domain can be directly viewed as an ER graph. RDFS extends the RDF symbols by new URIs for predefined class and relation types such as *rdfs:Resource* (the class of all resources), *rdfs:subClassOf* (for representing the *subclass* relationship), etc. RDFS is integrated into the more expressive Web Ontology Language.

OWL Going one step further, one can also associate assertions to each entity. These assertions are used to express claims about entities (e.g., humans cannot fly, humans are mortal, etc.). For this purpose, W3C recommends the Web Ontology Language (OWL) [53]. OWL allows the definition of domain resources and axioms about them. The axioms place constraints on entity classes and the types of relationships permitted between them. For example, axioms about persons might state that the relation *hasParent* can only be present between two persons when either *hasFather* or *hasMother* is also present, or that if person A is older than B, then A cannot stand in a *hasMother* or *hasFather* relationship with B. In full generality, such axioms can be used to assert special characteristics of relationships, to define the complement of classes, to express that two or more classes are disjoint, or to define boolean combinations of classes. In addition, they can be used to restrict the cardinality of classes. Hence, these axioms allow systems to infer additional information about the represented entities. For example, a person of blood type 0 cannot be the child of persons of blood type AB.

On the other hand, since these axioms are constraints, they entail a high complexity for reasoning tasks. The satisfiability problem for OWL classes, namely the problem whether there exists an instance of a given OWL class, is undecidable. For this reason OWL comes with three ascending levels of expressiveness: OWL Lite, OWL DL, and OWL Full. The semantics of OWL Lite and OWL DL can be described through a special Description Logic. The satisfiability problem is decidable in both cases (in EXPTIME for OWL Lite and in NEXPTIME for OWL DL [84]). The reasoning for the more expressive language OWL Full is undecidable; but OWL Full is the only OWL variant that is fully compatible with RDFS.

3.3 Storing and Querying Entity-Relationship Graphs

A common way for managing large ER graphs is by storing them in database systems. Such systems allow the management of the stored information by using system-dependent manipulation and query languages.

3.3.1 Relational, Object-Oriented, and XML Databases

RDBMS In order to overcome the drawbacks concerning the structural inflexibility of hierarchical and network databases, relational database systems were introduced. In a relational database, data is organized in relational tables, in which each record forms a row with predefined attributes in it. Relational Database Systems have been widely used in business applications. Their major task has been to perform customer-oriented on-line transaction and query processing. They usually adopt application-oriented ER models for the database design and support structured querying and management of the stored data through the Structured Query Language (SQL).

In a relational database, the edges of an ER graph can be stored as rows in one or multiple tables which can be queried through SQL. The following sample query asks for nodes a and b that have a common predecessor; the assumption is that the edges of the ER graph are stored in a single table with the schema `graph(sourceID, relation, destinationID)`.

```
SELECT destinationID AS a, destinationID AS b
FROM graph INNER JOIN graph USING (sourceID)
WHERE a != b
```

SQL builds on Relational Algebra [43]. It is important to note that reachability queries (i.e., whether two nodes in the graph are connected) cannot be expressed in SQL. The paradigm behind SQL is precise querying and efficient processing that respects the data consistency.

Despite several benefits concerning simple storage with ad-hoc and descriptive querying possibilities, this flat representation of data leaves the graph semantics to the user. On the other hand, given the simplistic semantics of ER graphs, relational databases are a favored option for their management.

ORDBMS/OODBMS Object-relational database systems rely on the capabilities of SQL, and extend the relational data model by including object orientation to deal with added data types (e.g., user defined types, row types, reference types, collection types, etc.). Special SQL-based query constructs are included to deal with these data types. This extension comes with an increased complexity when processing the added data types.

Object-oriented database systems are based on the object-oriented programming paradigm, where each entity is considered as an object. Data and code relating to an object are encapsulated into a single unit. Each object is assigned a set of variables (for its description), a set of messages (for communication purposes), and a set of methods (holding the code for messages). Objects with common properties can be grouped into an object class, and classes can be organized into class-subclass hierarchies. Such systems support several features of object-oriented programming such as inheritance, overriding and late-binding, extensibility, polymorphism, etc. Further features, such as computational completeness, persistence, concurrency, recovery and ad-hoc querying are directly adopted from relational database systems. In this setting, the entity nodes of an ER graph could be modeled as objects and the relations could be modeled as properties of these objects. The objects stored in an object-oriented database can be queried and manipulated through the object-oriented query language, OQL. In difference to SQL, OQL supports object referencing within tables and can perform mathematical computations within OQL statements. However, all this comes with an increased complexity. Ad-hoc querying (as in SQL)

3.3. Storing and Querying Entity-Relationship Graphs

is in some cases impossible, because it compromises the encapsulation paradigm of object-orientation. A sample OQL query that asks for German physicists who had the same academic advisor is given in the following.

```
SELECT DISTINCT STRUCT (P1 :   phys1.name, P2 :
  (SELECT phys2.name FROM phys2 IN GermanPhysicist
   WHERE !phys1.name.equals(phys2.name)
     AND phys1.getAdvisor().equals(phys2.getAdvisor())))
FROM phys1 IN GermanPhysicist
```

By using the late-binding mechanism of object orientation, OQL can dynamically determine the class of an object. This is similar to computing the transitive closure of the *subclass* relationship in an ER graph. However, general transitive closure queries (i.e., queries that ask for the transitive closure of a relationship) cannot be expressed in OQL.

XML In the world of semi-structured data, the Extensible Markup Language (XML) [46] is the main option for allowing information systems to encode, serialize, and share structured data, especially via the Internet. The interpretation of XML encoded data is completely left to the application that reads it. The tree-based data model of XML makes it easy to hierarchically organize information by delimiting pieces of data and representing them as nodes of a tree structure. Hence, the XML data model is a straight-forward generalization of the relational model. More precisely, a relational table can be viewed as the root node of an XML document, the tuples of the table can be viewed as the children of the root node, and their children are given by the attributes in the tuples.

In order to query and manipulate collections of XML data, W3C has developed XQuery [49] which uses XPath [48] expressions to address certain parts of XML documents. Consider an XML document that contains information about physicists. Assume that each physicist is represented by a node tagged PHYSICIST and that each PHYSICIST node has children nodes tagged with NAME, NATIONALITY, ADVISOR, etc. A high-level overview of a so-called FLWR XQuery expression asking for German physicists who had the same academic advisor is depicted in the following.

```
FOR $phys1 IN doc(``physicists.xml'')//PHYSICIST,
    $phys2 IN doc(``physicists.xml'')//PHYSICIST
WHERE $phys1/ADVISOR = $phys2/ADVISOR
  AND ends-with($phys1/NATIONALITY, `German')
  AND ends-with($phys2/NATIONALITY, `German')
  AND $phys1/NAME != $phys2/NAME
RETURN $phys1/NAME, $phys2/NAME
```

3.3. Storing and Querying Entity-Relationship Graphs

Notations of the form `A//B` are shorthand XPath notations that ask for any descendant node tagged B when descending from A in the XML tree structure. Similarly, notations of the form `A/B` ask for direct children of A that are tagged with B.

The above query example illustrates that XML and XQuery are geared for tree structures. With added modules such as ID/IDREF [46] (for establishing key/foreign key references between XML elements), XLink [50] (for adding hyperlinks between XML elements or XML documents) and XPointer [51] (for adding pointers to parts of an XML document), XML documents can be viewed as graphs. However, the current recommendations of XPath 2.0 and XQuery 1.0 do not support the navigation along XLinks [27].

3.3.2 Storing RDF data

Usually RDF triples are directly mapped onto relational tables. In general, there are two main strategies for doing that:

1. All triples are stored in a single table with generic attributes representing the *Subject*, the *Predicate*, and the *Object*.

2. Triples are grouped by their predicate name, and all triples with the same predicate name are stored in the same property table

The storage strategies are crucial for querying the RDF data. Different storage strategies favor different query types. While the second storage strategy is efficient on simple predicate-based triple lookup queries (i.e., for triples with the same predicate), the first strategy favors entity-based triple lookup (i.e., for triples with the same entity). Furthermore, by means of self-joins, the first storage strategy allows complex join queries between entities in a straight-forward way. However, the efficiency for these kind of tasks degrades in the presence of tens of millions of triples. Therefore, hybrid strategies such as the one used by Jena [147, 7] or Sesame [34, 122] cluster triples by predicate names, but based on predicates for the same entity class or for the same workload. A recent approach, coined RDF-3X [119], eliminates the need for physical fragmentation of the RDF graph into multiple tables. It shows that by creating smart and exhaustive indexes over a single, large table of RDF triples, join-style querying can be done very efficiently.

3.3.3 SPARQL

The standard query language for RDF data is SPARQL [54] (recursive acronym that stands for SPARQL Protocol and RDF Query Language). In January 2008, it became a W3C Recommendation. SPARQL queries are pattern matching queries on triples from an RDF data graph. A high-level representation of a SPARQL query has the form

```
SELECT ?variable1 ?variable2 ...
WHERE { pattern1. pattern2. ... }
```

where each pattern consists of a subject, a predicate, and an object, and each of these is either a variable, a URI or a literal. The query model is query-by-example style: the query specifies the known literals and leaves the unknowns as variables. Furthermore, all patterns represent conjunctive conditions (denoted by the dot between two patterns). Hence, variables that occur in multiple patterns imply joins. A SPARQL query processor needs to find all possible variable bindings that satisfy the given patterns and return the bindings from the projection clause to the application. The following sample query asks for German physicists that have the same academic advisor.

```
SELECT ?phys1 ?phys2
WHERE { ?phys1 type GermanPhysicist.
        ?phys2 type GermanPhysicist.
        ?phys1 hasAdvisor ?advisor.
        ?phys2 hasAdvisor ?advisor.
      }
```

More abstractly, a SPARQL query defines a subgraph matching task. In the above example, the query aims to find all entity nodes `?phys1` and `?phys2` that are connected to a node `?advisor` through an edge labeled `hasAdvisor`. The pattern matching semantics requires that all bindings of `?phys1`, `?phys2`, and `?advisor` be computed. Although for this kind of subgraph matching tasks, the SPARQL syntax is more intuitive than the SQL, OQL, or the XQuery syntax, as SQL, OQL, and XQuery, it lacks the power of expressing reachability or transitive closure queries over relationship labels. In fact, [22] shows that (for a given schema) the expressive power of SPARQL (as recommended by W3C) is equivalent to that of Relational Algebra.

3.4 Summary

In this chapter, we formally introduced the notion of ER graphs. We presented RDFS and OWL as two prominent ER-based schema languages for representing the resources of a domain of interest and reasoning about them. While OWL supports the definition of axioms about resources, and is more expressive than RDFS, it suffers from high complexity or even undecidability for reasoning problems.

We gave an overview of state-of-the-art techniques for storing and querying ER graphs. In relational database systems, the edges of an ER graph can be conveniently

3.4. Summary

mapped onto flat relational tables with generic attributes representing the source entity, the relation label, and the destination entity of an edge. SQL can be used to query the stored graphs, but the user has to be aware of the graph semantics encoded in the flat tables. Object-oriented and object-relational database systems offer a richer semantics for representing and querying ER graphs by borrowing concepts from object orientation, but this richness comes with increased complexity for querying and processing the stored data. XML with XLink and XPointer can represent ER graphs, but current XML query languages, such as XPath and XQuery are geared for tree structures and cannot deal with general, possibly dense graphs. SPARQL offers an intuitive semantics for subgraph matching tasks in RDF data, but like the previous query languages, it lacks the power to express reachability queries or queries asking for the transitive closure of (transitive) relations.

In general, the database research community has mainly emphasized the aspects of data consistency, precise query processing, and efficiency. We, on the other hand, envision knowledge bases with expressive search and ranking capabilities, and embedded knowledge discovery techniques, specifically geared for ER graph structures.

4 NAGA

"If music had been invented ten years ago along with the Web, we would all be playing one-string instruments and not making great music."

UDI MANBER

4.1 Overview

Our vision is the world's most comprehensive knowledge base derived from the Web. An important step towards this vision is the extraction and organization of information into explicit and unifying structures. Another important step is the design of search techniques that leverage these structures.

Consider a knowledge base that organizes information from the Web in a huge graph with entities (e.g., persons, locations, organizations, dates, etc.) as nodes and relationship instances or facts (e.g., (Max_Planck, hasWon, Nobel_Prize), (Max_Planck, bornIn, Kiel), etc.) as edges. Such a knowledge base would pave the way for new querying techniques that are simple and yet more expressive than those provided by standard keyword-based search engines. It would give us the opportunity to search for explicit knowledge rather than Web pages.

In this chapter, we propose NAGA (Not Another Google Answer), a new semantic search engine. NAGA builds on a knowledge base, which organizes information as a graph with typed nodes and edges, and consists of millions of entities and relationships extracted from Web-based corpora. A graph-based query language enables the formulation of queries with advanced semantic information. We introduce a novel scoring model, based on the principles of generative language models. Our model formalizes the notions of confidence, informativeness, and compactness and uses them to rank query results. We demonstrate NAGA's superior result quality over state-of-the-art search engines and question answering systems.

4.1.1 Motivation and Problem Statement

MOTIVATION The Web has become the prime source of information. Today's search engines index rich textual features of billions of Web pages and exploit the link structure between Web documents for the retrieval process. On top of that, they can return answers to user queries within milliseconds.

4.1. Overview

However, all major search engines are still keyword-based, which means that they are restricted to finding keywords in Web pages. This is fully sufficient for simple information needs, but highly inconvenient for more advanced queries where the keywords need to be interpreted as entities or relationships.

As a concrete example, suppose we want to learn about physicists who were born in the same year as Max Planck. Posing this query to Google (by using the keywords "physicist born in the same year as Max Planck") yields only pages about Max Planck himself, along with pages about the Max-Planck Society. We also posed this query to state-of-the-art question answering systems such as Yahoo! Answers [150], START [10, 79], True Knowledge [12], Wolfram Alpha [15], and Powerset [9]. None of these systems could answer it. In Chapter 1, Section 1.1, we already mentioned the main problems that current keyword search engines and question answering systems have with answering this kind of queries. In summary, for the keyword search engines, one can say that the keyword-based and page-oriented search paradigm is not powerful enough for such search tasks. State-of-the-art question answering systems are rather focused on understanding and answering simple question patterns, and are obviously overstrained with the above search task.

This example highlights the need for more explicit, unifying structures for the information on the Web. A knowledge base which could understand binary predicates, such as *isA*(*Max_Planck, Physicist*) or *bornInYear*(*Max_Planck, 1858*) would go a long way in addressing information needs such as the above. For example, the above query could be expressed as a conjunctive query akin to Datalog. Figure 3 depicts its graph-based representation.

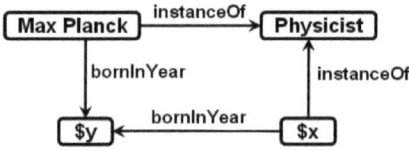

Figure 3: Example query

The nodes labeled with $x and $y represent variables that have to be bound by entities that satisfy the conjunctive conditions represented by the labeled edges of the query. The answer to such a query would be highly precise (by returning entities that satisfy the conditions).

Similar queries may often return hundreds, if not thousands, of results including uninteresting ones. Therefore, we need to think about ranking strategies that favor important results.

PROBLEM STATEMENT Consider a knowledge base that organizes information from the Web in an ER graph. The main problem that we address in this chapter is the design of a graph-based search framework that is intuitive and at the same time expressive enough to formulate queries with entities and relationships.

4.1. Overview

This problem comes with two main challenges:

1. Designing an expressive query language that allows formulating graph-based queries with entities and relationships in a convenient way.

2. Designing an answer and ranking model that prioritizes important and salient answers from the knowledge base.

PROBLEMS WITH PREVIOUS APPROACHES There are several research avenues that aim at this direction in a broader sense.

Graph-based querying of XML and RDF data or data mining on biological networks is a direction that is gaining momentum [80, 49, 31, 52, 54, 23], but does not consider the potential uncertainty of the data and disregards the need for a ranking model.

XML query languages such as XPath and XQuery [48, 49] have been extended to XQuery Full-Text [31, 52] with the purpose of ranked retrieval on semistructured data. Although this research direction considers ranking models, most of the proposed methods are geared for trees and do not carry over to richer knowledge graphs.

SPARQL [54], the query language for RDF data, is most similar to our query model, but it does not consider ranking and cannot express (transitive) connectivity queries or queries with regular expressions over predicate labels (e.g., to capture certain paths between entities).

Finally, entity-oriented (Web) search and other forms of "semantic" information retrieval [38, 41, 120] provide ranking but have rather simple query models for supporting keyword and record-level search.

Our work positions itself at the confluence of these research avenues and creates added value by combining techniques from all of them and further extending these synergetic approaches by various novel building blocks.

OUR APPROACH AT A GLANCE The data model of our semantic search engine, NAGA, builds on the ER-graph model. As introduced in Chapter 3, an ER graph is a labeled multi-graph. We call the labeled nodes of the graph *entities* and its labeled edges *facts*. Figure 1 (Chapter 2) depicts a sample from an ER graph. In that sample, the edge (*Max_Planck*, *fatherOf*, *Erwin_Planck*) represents a fact about the entities *Max_Planck* and *Erwin_Planck*.

In our data model, we assume that for each fact f we have all URLs of Web pages from which f was derived (i.e., pages from which f was extracted or in which f was recognized), and refer to these pages as the *witnesses* of f. We denote the set of witnesses of f by $W(f)$. Note that although there may be many witnesses for f, there is only one edge in the ER graph that represents f. From the witnesses, we compute for each fact f a *confidence weight*: $confidence(f)$. This weight depends on the estimated *accuracy* with which the fact f was derived from a witness p (denoted

4.1. Overview

by $accuracy(f, p)$), and the *trust* we have in p (denoted by $trust(p)$). The value $accuracy(f, p)$ is usually provided by the mechanism that is responsible for deriving f from p. The trust $trust(p)$ in p can be computed by any algorithm similar to PageRank. With these ingredients, the confidence of f can be computed as:

$$confidence(f) = \max\{accuracy(f,p) \times trust(p) | p \in W(f)\} \qquad (4.1)$$

This is only one way (among various options) of combining the above aspects to a confidence value. The assumption behind Equation (4.1) is that pages with high trust (i.e., high authority) are used as primary sources for information extraction, as they are likely to contain accurate and clean information. In such a setting, where there are many pages that have a similarly high trust, the extraction accuracy should be the critical factor. These confidence weights are used in NAGA's ranking model.

In order to query the knowledge-graph, NAGA provides a graph-based query language that supports queries about entities and relationships. This queries can be simple conjunctive queries similar to the one depicted in Figure 3, but they can also be more complex by exploiting regular expressions over relationships as edge labels. Figure 4 depicts a sample query that asks for philosophers from Germany who have been influenced by the English philosopher William of Ockham.

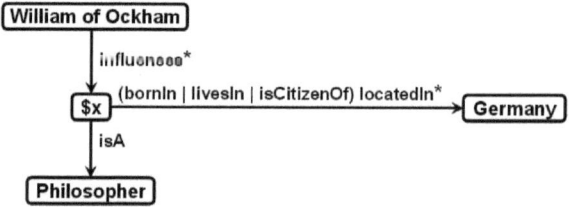

Figure 4: Example for a complex query

Without going into details, answers to NAGA queries are given by subgraphs (of the underlying ER graph) that match the structure, the entity names as well as the relationship expressions of the query graph. Whenever a query yields multiple answers, NAGA ranks them by means of a novel scoring mechanism that is based on the principles of generative language models for document-level information retrieval [115, 152]. We apply these principles to the specific and unexplored setting of weighted, labeled graphs. Our scoring model is extensible and tunable and considers several intuitive notions like *compactness*, *informativeness*, and *confidence* of results.

As of now, NAGA operates on the YAGO knowledge base [137, 138, 136]. YAGO contains more than 20 million facts about approximately 2 million entities. It combines facts extracted from semi-structured Wikipedia sources with facts from the WordNet thesaurus [16, 72] (see Chapter 2, Section 2.2). NAGA operates on more than 100 predefined relationship labels provided by YAGO such as *isA*, *means*, *bornOnDate*, *hasChild*, *isMarriedTo*, *establishedOnDate*, *hasWonPrize*, *locatedIn*, *politicianOf*, *graduatedFrom*, *actedIn*, *discovered*, *isCitizenOf*, etc.

4.1. Overview

4.1.2 Related Work

INFORMATION EXTRACTION AND ONTOLOGY BUILDING Approaches that aim at information extraction, e.g., [19, 67, 25, 55, 68, 91, 130, 137, 146], and ontology building, e.g., [56, 4, 24, 127, 136], are in a broader sense related to our work, as they are driven by the vision of semantic information processing on the Web. However, they do not primarily address querying and ranking models for the acquired knowledge. TextRunner [67], for example, provides a query interface for simple record-oriented search [11]. More elaborated are the query interfaces of DBpedia [24] and freebase [4]. DBpedia offers a SPARQL query endpoint, and also supports queries asking for broad connections between two given entities [109]. For the casual user, freebase provides an interface for keyword queries, and for the experienced users it also supports structured queries. Neither DBpedia nor freebase consider the potential uncertainty of the data, and they both disregard the need for ranking. Finally, YAGO [136] uses NAGA for query purposes.

ENTITY AND RECORD SEARCH Prominent approaches addressing the problem of entity-centric search are Libra [120], Cimple [59, 134], ExDBMS [35], and EntityRank [41, 42]. All these approaches operate on entity-centric records extracted from Web data.

Libra is part of a comprehensive technology for information extraction and entity-oriented search. Pattern-matching algorithms that are tailored to typical Web-page layouts are trained (by means of advanced models like Hierarchical Conditional Random Fields [153]) to learn patterns for extracting entities and their attributes from product-related pages with HTML tables and lists. The goal is to build and maintain several vertical-domain portals, including product search and the Libra portal for scholarly search on the extracted records about authors, papers, conferences, and communities. Once the facts are gathered and organized into a searchable form, they can be queried with Libra. In order to rank results, Libra uses an advanced statistical language model, extended from the level of document-oriented bags-of-words to structured records. However, Libra does not consider general relations between different entities, and its query model is keyword-based.

Similarly to Libra, the Cimple project [59, 134] aims to generate and maintain searchable, community-specific portals with structured information gathered from Web sources. Cimple's flagship application is the DBLife portal [2]. DBLife features automatically compiled "super-homepages" of researchers with bibliographic data as well as facts about community services (PC work, etc.), colloquium lectures, and more. A suite of extractors (that build on pattern matching and dictionary lookups) are periodically combined into execution plans to extract facts from a carefully selected set of relevant Web sources (e.g., DBLP, Dbworld, university pages, etc.). To query the gathered data about entities, a Datalog-based query-language can be used. Database rewriting techniques are exploited for query optimization (see [89]). However, the problem of ranking for the returned answers is not addressed.

4.1. Overview

EntityRank [41, 42] facilitates search that can combine keywords and structured attributes in a convenient manner. The authors view the Web as a repository of entities and address the problem of answering user queries that are composed of keywords and entities. The answers contain explicit entity instances corresponding to the entities in the query. For example, when a user query contains the entity 'email' the answer should contain email addresses. In order to rank the entity instances, the authors introduce a probabilistic ranking model. The model aims to formalize the impression a user (who has no resource or time constraints) would have when he repeatedly visits all Web pages that contain instances of the query entities. Despite its elaborated ranking framework, EntityRank does not address typed relations between entities and its query language builds on the keyword search paradigm.

ExDBMS [35] exploits a suite of powerful extractors (e.g., TextRunner [11], KnowItAll [68], Dirt [114], etc.) to build a database with facts extracted from Web pages. To capture the uncertainty of the extracted data, the facts are assigned probability values. In order to query the extracted data, ExDBMS supports a probabilistic form of Datalog [58]. The returned facts are ranked by their probability values. In contrast, our engine, NAGA, uses a graph-based search paradigm that is more expressive by supporting regular expressions over relationships and broad connectivity queries. Furthermore, NAGA extends all above approaches by adding to the data uncertainty aspect of ranking further important aspects that capture the importance and the succinctness of results.

There are certainly many other approaches which address the problem of entity and record search, especially in the areas of Deep-Web search, vertical search, and semantic desktop search (e.g., [38, 62, 63, 116]). Usually, they aim at enhancing keyword-based querying by typed attributes, but none of these approaches is sufficiently complete for effectively searching a richly structured knowledge base.

QUESTION ANSWERING SYSTEMS The ambitious goal of automatically answering questions posed in natural language has been addressed by various systems. Prominent ones are Wolfram Alpha [15], Answers [1], Powerset [9], Hakia [6], True Knowledge [12] Yahoo! Answers [150], and START [10, 79]. Most of them employ Natural Language Processing techniques to parse and understand the posed questions. Wolfram Alpha, Answers, Powerset, Hakia, and True Knowledge are commercial, and very little is known about the technology used in the background. However their results hint at structured and well-annotated data used in the background for answering questions.

Yahoo! Answers is a commercial system as well. But in contrast to the other approaches, it builds on the "wisdom-of-the-crowds" paradigm. It has its own corpus of questions posed by users and corresponding answers given by users. For every posed question, the system first tries to match it to a question from the corpus. When a match is found the corresponding answer is returned. Otherwise the question is presented as an unanswered question to the user community. After the question has been answered, both, the question and the answer are added to the corpus.

4.1. Overview

START is an established Web-based question answering system. It has been developed by the InfoLab Group at the MIT Computer Science and Artificial Intelligence Laboratory. START exploits information extracted from various Web sources as a background corpus. Its main knowledge source is Wikipedia. In contrast to NAGA's fact-based knowledge base, START uses whole text chunks as well as information extracted from Wikipedia lists. In a natural-language-processing phase, the query is analyzed and its focus is identified by determining the main concepts in the query. WordNet is exploited to identify hyponyms of the main concepts from the query. Finally, the most promising text snippets that contain these hyponyms are identified in the background corpus and returned as answers. The snippets are ranked based on structural analysis and $tf * idf$-based scores with respect to the main terms from the question.

True Knowledge seems to follow a strategy similar to that of START. It attempts to comprehend posed questions by first identifying their most likely meaning. A knowledge base with explicit facts about entities is used to answer user questions. As reported on [13], the system can reason about the facts in its knowledge base. However, there is no information about how this reasoning is exploited to retrieve or rank answers.

Wolfram Alpha was released to the public on May 15, 2009. The answers to user queries are computed from structured data and supported with comprehensive visualizations. Wolfram Alpha performs surprisingly well on mathematical questions. The answer usually presents a human-readable solution. This distinguishes Wolfram Alpha from many semantic search engines.

All question answering engines presented above have often problems understanding or dealing with questions for which the answer has to be composed from different pieces of information distributed across multiple Web pages. For example, none of the engines could answer the question about physicists who were born in the same year as Max Planck.

GRAPH-BASED SEARCH AND INFORMATION RETRIEVAL The need for querying semistructured and RDF data has led to query languages such as XPath or XQuery [48, 49] (for XML data), SPARQL [54] and extensions [23] (for RDF data). However, the proposed query languages disregard the issue of uncertainty and ranking, and are often not expressive enough to capture transitive relations or broad connections between entities.

Another research area, related to our work, addresses the problem of ranked retrieval in semi-structured data (see [21] and the references given there). Researchers from this area have proposed query languages that combine variations of XQuery constructs with full-text search. We give some examples in the following.

XXL [132] deals with querying hyperlinked XML documents (i.e., graph structures). Its query language supports path expressions combined with similarity search for terms. Based on an inductively defined relevance score (i.e., $tf * idf$-term scores and ontology-based similarity scores for XML elements are combined to

4.1. Overview

relevance scores for sub-graphs), the answer to a query is defined as a ranked list of XML sub-graphs which match the graph structure of the query. The latter is similar to the query answering approach of our work. Nevertheless, our query language is more powerful by allowing search for regular expressions over general relationships and for connectivity between entities. This makes a big difference in the match and ranking semantics.

SphereSearch [78, 77] casts Web pages and the links between them into an XML graph. Its query language supports similarity-aware search by combining keywords with entity classes (i.e., concepts) and attributes. The query answering model builds on the idea that closely interlinked Web pages may contain logically related information (i.e., the idea of *information unit* [112]). The results are compact subgraphs of the XML graph which capture the context of the query (as given by the keywords and concepts). In contrast to the framework of SphereSearch, our framework is general enough to capture logically related information from pages that are not interlinked.

XSEarch [45] proposes a novel tree-based interconnection semantics for XML elements. Its query language is keyword-based. For a given query, XSEarch exploits the above semantics to retrieve XML subtrees the nodes of which contain the query keywords. In order to rank results, XSEarch applies the notions of $tf * idf$ and "cosine similarity" to the setting of XML trees. The approach of XSEarch is improved in [44] in two ways: (1) the underlying structure is generalized to a graph structure by taking ID references into account, (2) a document schema is exploited for computing answers.

Finally, there is prior work on keyword proximity search in schema-oblivious database graphs. The graphs are usually obtained by viewing the tuples of database tables (or the tables themselves) as nodes and the foreign-key relationships between tuples (or relations) as edges. These kinds of data graphs can be generalized into networks of entities and relationships, and similar graph structures also arise when considering XML data with XLinks and other cross-references within and across document boundaries [44, 78]. In this setting, a query consists of keywords, and a node of the graph contains a keyword if the corresponding tuple (or relation) contains it. For a given query, the goal is to determine the smallest subgraph that interconnects the nodes containing the keywords. By taking node or edge weights into account, the problem becomes NP-hard. Hence, prominent systems such as BANKS [28, 92], BLINKS [82], DBXplorer [20], and DISCOVER [86] solve this problem heuristically. Remarkable are also the approximation guarantees as well as the efficiency results of [61, 101, 131]. We will take a detailed look at these approaches in Chapter 5.

All approaches presented above cover important issues with respect to graph search and ranked retrieval on graphs. However, none of them provides a holistic search and ranking model that exploits the inherent semantics of entities and explicit relationships in ER graphs. NAGA instead, makes the explicit nature of ER graphs a key ingredient of its search framework.

4.1.3 Contributions and Outline

Our search engine, NAGA, provides a novel and holistic framework for graph-based search with entities and relationships.

Our major contributions in this chapter are:

1. An expressive and concise query language for searching a Web-derived knowledge base.

2. A novel ranking model based on a generative language model for queries on weighted and labeled graphs.

3. An extensive evaluation of the search-result quality provided by NAGA, based on user assessments and in comparison to state-of-the-art search engines and question answering systems like Google, Yahoo! Answers, and START [79]. Furthermore, we demonstrate the superiority of NAGA's ranking mechanism over comparable mechanisms as used in [28, 92].

The rest of this chapter is organized as follows. In Section 4.2, we present NAGA's query and answer model. In Section 4.3, we describe NAGA's ranking model. The architecture and the implementation details of the NAGA engine are presented in Section 4.4. Section 4.5 is devoted to the experimental evaluation of the NAGA system. We conclude in Section 4.6.

4.2 A Framework for Querying with Entities and Relationships

4.2.1 Query Model

NAGA's query model is derived from the definition of ER graphs. As in Chapter 3, let Ent and Rel be finite sets of entity and relationship labels, respectively, and let $G = (V, l_{Ent}, E_{Rel})$ be the ER graph representing the underlying knowledge base. We denote by $RegEx(Rel)$ the set of regular expressions over Rel, and by $\mathcal{L}(r)$ ($\subseteq Rel^*$) the language of a regular expression $r \in RegEx(Rel)$.

DEFINITION 2: [*NAGA Query*]
Let Var be a set of variables, such that $Var \cap Ent = \emptyset$ and $Var \cap Rel = \emptyset$. A NAGA query over Ent, Rel and Var is a connected directed graph $Q = (V_Q, l^Q_{Ent}, E^Q_{Rel})$, where V_Q is a finite set of nodes with $V_Q \cap V = \emptyset$, $l^Q_{Ent} : V_Q \rightarrow Ent \cup Var$ is a function that maps query nodes to entity labels or variables, and $E^Q_{Rel} \subseteq (Ent \cup Var) \times (RegEx(Rel) \cup Var) \times (Ent \cup Var)$ is a finite set of labeled edges.

We call a node or an edge labeled with a variable *unbound*. Variables are placeholders for entity or relationship labels.

4.2. A Framework for Querying with Entities and Relationships

As in the definition of ER graphs, the labeled nodes stand for entities and the labeled edges stand for relationship instances or facts.

Given a NAGA query $Q = (V_Q, l^Q_{Ent}, E^Q_{Rel})$, we call a triple $f = (x, r, y) \in E^Q_{Rel}$ (i.e., a query edge) a *fact template*. For example, (Albert_Einstein, instanceOf subclass*, $x) is a fact template. Here, $x denotes a variable, and *instanceOf subclass*** is a regular expression over relationship labels. The template asks for all classes Albert Einstein belongs to (e.g., physicist, philosopher, scientist, person, entity, etc.). The exact query semantics is described in our answer model.

4.2.2 Answer Model

NAGA's answer model is based on subgraph matching. As before, let $G = (V, l_{Ent}, E_{Rel})$ denote the ER graph of our knowledge base. For a given query, NAGA aims to find subgraphs of G that match the query graph.

We say that a node $v \in V$ *matches* a query node with label λ, if $l_{Ent}(v) = \lambda$ or if λ is a variable. Furthermore, we say that a query node $v' \in V_Q$ is *bound* to a node v of G if v matches v'.

In the following, for a labeled edge (i.e., a fact) $f = (\alpha, \beta, \gamma)$ of G, we refer to its relationship label β by $rel(f)$. Note that $\alpha, \gamma \in Ent$ and $\beta \in Rel$. Before defining matches to NAGA queries, we define matches to fact templates.

DEFINITION 3: [*Matching Path*]
Let the wildcard ".*" denote the regular expression over Rel that stands for any sequence of relationship labels. A matching path for a fact template (x, r, y) is a sequence of labeled edges $m_1, \ldots, m_n$ from G, such that the following conditions hold:

- If r is a variable, then $n = 1$ and the start node of m_1 matches x and the end node of m_1 matches y.

- If r is a regular expression different from the wildcard ".*", then $m_1, \ldots, m_n$ forms a directed path and $rel(m_1) \ldots rel(m_n) \in \mathcal{L}(r)$. Furthermore, the start node of m_1 matches x and the end node of m_n matches y.

- If $r = .*$, then $m_1, \ldots, m_n$ forms an undirected path, such that its start node matches x and its end node matches y.

The direction of a relationship label on an edge is associated with the direction of the edge (i.e., the direction of the edge reflects the *subject-predicate-object* order). In our definition, when the regular expression of a fact template is different from ".*", we assume the same *subject-predicate-object* order for each relationship label occurring in the regular expression and require that the matching path be directed.

4.2. A Framework for Querying with Entities and Relationships

When a query edge is labeled with ".*", we are interested in a broad connection between the two nodes of the edge. Hence, we drop the requirement of directed paths.

In the following, we generalize the *match* definition to queries.

DEFINITION 4: [*Answer Graph*]
An answer graph *to a query q is a subgraph S of G, for which the following conditions hold*:

1. *For each fact template in q there is exactly one matching path in S.*

2. *Each fact in S is part of a matching path.*

3. *Each node of q is bound to exactly one node of S.*

For a query q with query templates $q_1, \ldots, q_n$ and an answer graph g, we denote the matching path of a query template q_i from q by $match(q_i, g)$.

We will use the label *isA* as a shorthand for the regular expression *instanceOf subclass**. The expression *isA* connects an individual via one *instanceOf*-labeled edge to its immediate class and by several *subclass*-labeled edges to more general superclasses.

NAGA provides two query types associated with different levels of expressiveness: (1) *simple-relationship queries* and (2) *regular-expression queries*.

4.2.3 Simple-Relationship Queries

Simple-relationship queries are in the spirit of SPARQL or conjunctive Datalog queries.

Formally, a *simple-relationship query* is a NAGA query $Q = (V_Q, l_{Ent}^Q, E_{Rel}^Q)$ in which for every fact template $(x, r, y) \in E_{Rel}^Q$, we have that $r \in Rel \cup Var$.

The query from Subsection 4.1.1 that asks for *physicists who were born in the same year as Max Planck* (see Figure 3) is an example for such a query. Further examples are depicted below.

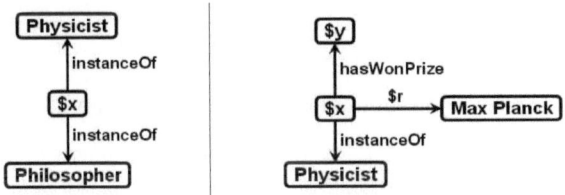

Figure 5: Examples of simple-relationship queries

The query on the left asks for *philosophers who are also physicists*. The query on the right asks for *prizes won by physicists related to Max Planck*.

4.2. A Framework for Querying with Entities and Relationships

In order to compute answers to these queries, NAGA attempts to bind the variables by finding a subgraph from the knowledge base that matches the query. Figure 6 depicts two results to the above queries as returned by NAGA. Note that for each of the above queries there are multiple answers, and NAGA returns a ranked list of answers. The depicted results are both the top-ranked answers. The answer graph on the left contains Aristotle, one of the most influential philosophers who was also a physicist. Further influential physicists and philosophers like Albert Einstein, David Bohm, and Anaxagoras can be found in the top-10 results returned by NAGA. The answer on the right contains Max von Laue who was a student of Max Planck and won the Nobel Prize for the discovery of X-ray diffraction by crystals, an important method for analyzing atomic structures. How NAGA ranks the results will be explained in detail in Section 4.3. Next, we give an overview of regular-expression queries.

Figure 6: Answers to example queries of Figure 5

4.2.4 Regular-Expression Queries

Regular-expression queries give users the flexibility to express and capture vague or transitive relations between entities.

Formally, a *regular-expression query* is a NAGA query $Q = (V_Q, l^Q_{Ent}, E^Q_{Rel})$ in which there is at least one fact template $(x, r, y) \in E^Q_{Rel}$ with $r \in RegEx(Rel) \cup Var$.

Note that every simple-relationship query is a regular-expression query, but not vice versa.

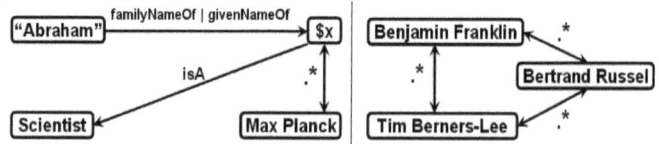

Figure 7: Regular-expression query examples

The query from Subsection 4.1.1 that asks for *philosophers from Germany who have been influenced by the English philosopher William of Ockham* (see Figure 4) is an example for a regular-expression query. In that query, we used the fact template ($x, (bornIn|livesIn|isCitizenOf) locatedIn*, Germany$) to express that we are asking for a philosopher *from Germany*. NAGA returns Albert Einstein (who was U.S.-Swiss citizen of German origin) and Goethe as top results.

Further examples are depicted below.

Suppose that you heard about a scientist named "Abraham" in connection with Max Planck. The query on the left (Figure 7) asks for a scientist by the name of "Abraham" and his connection to Max Planck. NAGA says that Max Planck was the academic advisor of the physicist Max Abraham. Note that in this query, the relationship label *isA* is a short-hand notation for the regular expression *isntanceOf subclass**.

The query on the right asks for a broad relation between Tim Berners-Lee, Benjamin Franklin, and Bertrand Russel. NAGA tells us that all three of them are fellows of the Royal Society. It turns out that from an efficiency viewpoint, these queries are very hard to answer. In Chapter 5, we will present our algorithmic solution for retrieving answers to these kinds of queries.

4.3 A Framework for Ranking with Entities and Relationships

Designing ranking models for ER subgraphs is a challenging task. The ranking criteria should comply with the human intuition about important results.

4.3.1 Ranking Desiderata

We think that a good ranking model for answer graphs should satisfy the following desiderata:

1. *Confident* answers (i.e., answers containing facts with high extraction confidence from authoritative pages) should be ranked higher.

2. *Informative* answers should be ranked higher. For example, when asking the query (*Albert_Einstein, isA, $z*) the answer (*Albert_Einstein, isA, Physicist*) should rank higher than the answers (*Albert_Einstein, isA, Philosopher*) or (*Albert_Einstein, isA, Person*), because Einstein is rather known as a physicist than as a philosopher, and the fact that Einstein is a person is rather trivial. Similarly, for a query such as ($y, isA, Physicist$), the answers about world-class physicists should rank higher than those about hobby physicists.

3. *Compact* answers should be favored, i.e., direct connections should be preferred to loose connections between entities. For example, for the query "How are Einstein and Bohr related?" the answer about both having won the Nobel Prize should rank higher than the answer that Tom Cruise connects Einstein and Bohr by being a vegetarian like Einstein, and by being born in the year in which Bohr died.

4.3. A Framework for Ranking with Entities and Relationships

We propose a novel ranking model that integrates all the above desiderata in a unified framework. Our approach is inspired by existing work on language models (LM) for information retrieval (IR) on document collections [152, 83], but it is adapted and extended to the new domain of knowledge graphs. In this setting, the basic units are not *words*, but *facts* or *fact templates*. Our graphs and queries can be seen as sets of facts or fact templates respectively. A candidate result graph in our setting corresponds to a document in the standard IR setting.

The language model we envision is much more challenging than the traditional language models for two reasons:

1. By considering facts and fact templates as IR units, rather than words in documents, our queries include both bound and unbound arguments – a situation that is very different from what we encounter in multi-term queries on documents.

2. Our corpus, the knowledge graph, is virtually free of redundancy (each fact occurs only once), unlike a document-level corpus. This makes reasoning about background models and idf-style aspects [152] more subtle and difficult.

4.3.2 Statistical Language Models for Document Retrieval

A critical issue for keyword search engines is the design of an effective retrieval model that can rank documents with respect to a given query. This has been a central research problem in information retrieval for several decades. An important group of ranking models are the statistical language models [117, 126, 115, 152, 108, 83] which have been successfully applied to many document-centric retrieval problems.

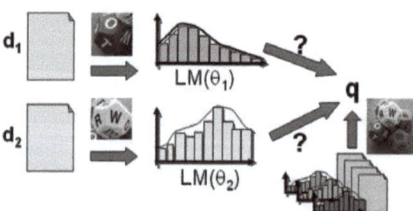

Figure 8: Idea of Language Models for Document Retrieval (source: [144])

As depicted in Figure 8, the basic idea is that each document d has its own language model (LM): a probability distribution over words with parameters Θ_d. Querying is viewed as a generative process. More precisely, for a given a keyword query $q = q_1 \ldots q_m$ and a document d, the query q is viewed as a sample from d. The task is to estimate the likelihood that the keywords of q were generated by the LM of d (i.e., that q is a sample of the LM of d). The documents are then ranked based on the probability of their language model having generated the given query.

4.3. A Framework for Ranking with Entities and Relationships

The score of d with respect to q is computed as:

$$score(d,q) = P(q|d) = P(q|\Theta_d) = P(q_1 \ldots q_m|\Theta_d) \approx \prod_{i}^{m} P(q_i|\Theta_d) \quad (4.2)$$

The last approximation in Equation (4.2) is based on the conditional independence assumption between the query terms given the parameters Θ_d. The independence assumption is widely used in information retrieval to avoid high computational complexity and sparseness problems in high-dimensional data settings ("the curse of dimensionality"). More specifically, $P(q_1 \ldots q_m|\Theta_d)$ could only be estimated if there were enough examples for all possible word sequences of length m in the documents. In reality, the data is very sparse in terms of such examples. Hence, for the maximum likelihood estimation of $P(q|\Theta_d)$ the conditional independence assumption between the query terms is made.

More advanced models, such as [83, 152], postulate conditional independence between the query terms and compute the score of a document d with respect to a query q as:

$$score(d,q) = P(q_1 \ldots q_m|\Theta_d) = \prod_{i=1}^{m} (\lambda_i P(q_i|\Theta_d) + (1-\lambda_i) P(q_i)) \quad (4.3)$$

where $P(q_i)$ is the probability of drawing the term q_i randomly from the underlying corpus (collection of all documents), $P(q_i|\Theta_d)$ is the probability of drawing a term randomly from the document d, and λ_i is a relevance weight for the term q_i. In this probabilistic mixture model, the term $P(q_i)$ corresponds to a background model which is responsible for the smoothing (i.e., for encountering the noise in the data). For example, if a term q_i from a query q is not contained in a document d, the estimation of Equation (4.2) will return $score(d,q) = 0$. The smoothing component of Equation (4.3), given by $(1-\lambda_i)P(q_i)$, avoids this by taking the occurrences of q_i in the whole corpus into account.

The above background model is similar to the idf component in the $tf * idf$ measure. In fact, the whole estimation in Equation (4.3) can be cast into a $tf * idf$-style measure. This can be best seen in the following transformations. We start with Equation (4.3), i.e.,

$$P(q_1 \ldots q_m|\Theta_d) = \prod_{i=1}^{m} (\lambda_i P(q_i|\Theta_d) + (1-\lambda_i) P(q_i))$$

Dividing the above formula by $\prod_{i=1}^{m}((1-\lambda_i P(q_i)))$ will not affect the ranking, because λ_i and $P(q_i)$ depend only on the query and have the same value for each document. Then, we have the rank equivalence:

$$P(q_1 \ldots q_m|\Theta_d) \propto \prod_{i=1}^{m} \left(1 + \frac{\lambda_i P(q_i|\Theta_d)}{(1-\lambda_i) P(q_i)}\right) \quad (4.4)$$

4.3. A Framework for Ranking with Entities and Relationships

Now, the term $P(q_i|\Theta_d)$ corresponds to the frequency of q_i in d and the term $P(q_i)$ corresponds to the document frequency of q_i. Hence, we have here an analogy to the $tf * idf$-style measures.

In the next subsection, we will see how these models can be applied to the previously unexplored setting of facts and fact templates from ER graphs.

4.3.3 The NAGA Ranking

In line with the models presented in [83, 152], we assume that a NAGA query q with fact templates $q_1 \ldots q_m$ is generated by a probabilistic model based on a candidate result graph g consisting of the facts $g_1 \ldots g_n$, $n >= m$. Our goal is to estimate the conditional probability $P(q|g)$, i.e., the probability that g generated the observed query q [152].

Assuming conditional independence between the query's fact templates given the candidate result graph g results in:

$$P(q|g) = P(q_i \ldots q_m|g) = \prod_{i=1}^{m} P(q_i|g) \qquad (4.5)$$

Our intuition behind the independence assumption is based on the independent extraction of facts in the construction phase of NAGA's knowledge base (see [138]). Furthermore, as discussed above, the independence assumption helps avoiding sparseness and intractability problems.

Next, we design a $tf * idf$-style probabilistic mixture model for fact templates. We follow classical IR literature [83] but develop a new scoring model suited for our setting.

We define the likelihood of a query template, given an answer graph, as a mixture of two distributions, $\tilde{P}(q_i|g)$ and $\tilde{P}(q_i)$ as follows:

$$P(q_i|g) = \alpha \cdot \tilde{P}(q_i|g) + (1-\alpha) \cdot \tilde{P}(q_i), \ 0 \leq \alpha \leq 1 \qquad (4.6)$$

$\tilde{P}(q_i|g)$ is the probability of drawing q_i randomly from an answer graph, $\tilde{P}(q_i)$ is the probability of drawing q_i randomly from the total ER graph and α is either automatically learned (via EM iterations [83]) or set to an empirically calibrated global value. Note that the value $\tilde{P}(q_i)$ is the same for all answers. As discussed in the previous subsection, there is a connection between this style of probabilistic models and the popular $tf * idf$ measure.

Our goal is to capture the three desiderata presented in Subsection 4.3.1: *confidence*, *informativeness*, and *compactness*.

We first describe the confidence and informativeness components and then explain how our model automatically deals with compactness. We describe $\tilde{P}(q_i|g)$ by a mixture model which puts different weights on confidence and informativeness. This is close in spirit to linear interpolation models used for smoothing [152]. The weight β is empirically calibrated as analyzed in our evaluation section.

$$\tilde{P}(q_i|g) = \beta \cdot P_{conf}(q_i|g) + (1-\beta) \cdot P_{info}(q_i|g), \ 0 \leq \beta \leq 1 \qquad (4.7)$$

4.3. A Framework for Ranking with Entities and Relationships

Note that the confidence and the informativeness are indeed independent criteria. For example, we can be very confident that Albert Einstein was both a physicist and a politician, but the former fact is more informative than the latter, because Einstein was a physicist to a larger extent than he was a politician.

ESTIMATING CONFIDENCE The maximum likelihood estimator for $P_{conf}(q_i|g)$ is:

$$P_{conf}(q_i|g) = \prod_{f \in match(q_i,g)} P(f \text{ holds}) \qquad (4.8)$$

where $P(f \text{ holds})$ is estimated as in Equation (4.1) by the confidence of f:

$$confidence(f) = \max\{accuracy(f,p) \times trust(p) | p \in W(f)\}$$

$W(f)$ denotes the witnesses (i.e., Web pages) from which f was derived, $accuracy(f,p)$ represents an accuracy with which f was derived from p, and $trust(p)$ captures the trust we have in p.

In case q_i is labeled with a label from $Rel \cup Var$, then $match(q_i, g)$ contains just one fact and $P_{conf}(q_i|g)$ is the confidence of that fact. If q_i is labeled with a regular expression over relations, then $match(q_i, g)$ contains the sequence of facts that together match q_i. The combined confidence then is the product of the confidences of the single facts in the sequence – assuming that the facts are independent.

ESTIMATING INFORMATIVENESS In the following (for simpler notation and ease of explanation), for a given query q and an answer g with facts $g_1 \ldots g_n$, we assume that q consists as well of n fact templates $q_1 \ldots q_n$ and that each template q_i is matched by the fact g_i in g.

The *informativeness* of a query template q_i given the answer graph g depends on the informativeness of the fact that matches q_i in g. As in our assumption, let g_i be the match of q_i in g. We estimate $P_{info}(q_i|g)$ as:

$$P_{info}(q_i|g) = \frac{|W(g_i)|}{|W(q_i)|} \qquad (4.9)$$

where $|W(g_i)|$ and $|W(q_i)|$ denote the number of witness pages for the fact g_i and the template q_i, respectively. We compute the number of witnesses for a given fact template by summing up over the number of witnesses for the facts that match the template. For example, consider the fact template q_i=(x, *instanceOf*, *Physicist*). We compute the number of witnesses for the template q_i as:

$$|W(q_i)| = \sum_x |W(x, \textit{instanceOf}, \textit{Physicist})| \qquad (4.10)$$

where x stands for any entity that occurs in an *instanceOf* relationship with the entity *Physicist*.

4.3. A Framework for Ranking with Entities and Relationships

In full generality, the witnesses could also be weighted by their authority (e.g., PageRank).

To see why the above formulation captures the intuitive understanding of infor-mativeness, consider the following examples. Let q be the query $q = $ (Albert_Einstein, instanceOf, x), which consists of one fact template. Let $f = $ (Albert_Einstein, instanceOf, Physicist) be a possible answer. Here, the informativeness measures how often Einstein is mentioned as a physicist as compared to how often he is mentioned with other instanceOf facts. Thus, $f = $ (Albert_Einstein, instanceOf, Physicist) will rank higher than $f' = $ (Albert_Einstein, instanceOf, Politician). In this case, informativeness measures the *degree* to which Einstein was a physicist.

Now consider the query $q = $ (x, instanceOf, Physicist) and consider again the answer $f = $ (Albert_Einstein, instanceOf, Physicist). In this case, the informativeness will compute how often Einstein is mentioned as a physicist as compared to how often other people are mentioned as physicists. Since Einstein is an important individual among the physicists, (Albert_Einstein, instanceOf, Physicist) will rank higher than (Bob_Unknown, instanceOf, Physicist). In this case, informativeness measures the *importance* of Einstein in the world of physicists.

More examples could be: when asking for prizes that Einstein won, our informativeness will favor the prizes he is most known for; when asking for people born in some year, informativeness favors famous people; when asking for the relationship between two individuals, informativeness favors the most prominent relation between them, etc.

For the currently compiled YAGO knowledge base, the number of witnesses for each fact is not statistically significant, because our facts are extracted only from a limited number of Web-based corpora, and many facts appear only on one page. For this reason we approximated the numbers of witnesses by the following heuristics. We transform the facts into keyword queries and use a search engine to retrieve the number of pages in the Web that contain the corresponding keywords. For example, to estimate $|W(\text{Albert\_Einstein}, \text{instanceOf}, \text{Physicist})|$, we formulate the query "Albert Einstein" + "physicist" and retrieve the number of hits for this query. Analogously, to estimate $\sum_x |W(x, \text{instanceOf}, \text{Physicist})|$, we retrieve the number of hits for the query "physicist". The reason for omitting the relationship label is that relationships are often expressed in non-trivial ways, which makes it impossible to capture them by means of keywords. To conclude the example, for the query (x, instanceOf, Physicist) and the answer (Albert_Einstein, instanceOf, Physicist), we estimate the informativeness as:

$$\frac{|W(\text{Albert\_Einstein}, \text{instanceOf}, \text{Physicist})|}{\sum_x |W(x, \text{instanceOf}, \text{Physicist})|} \sim \frac{\#hits(Albert\ Einstein\ physicist)}{\#hits(physicist)} \qquad (4.11)$$

In the evaluation section, we will see that in practice this approximation leads to a

4.3. A Framework for Ranking with Entities and Relationships

nice ranking behavior.

An alternative idea for computing the informativeness of facts is to exploit the structure of the underlying ER graph. More precisely, based on the endorsement hypothesis, one could estimate informativeness by taking the in-degree of nodes into account. The higher the in-degree of a node, the higher should be the authority of the corresponding entity. However, there are several problems with this approach. First, the direction of an edge in an ER graph does not necessarily reflect an endorsement. For example, the fact (*Albert_Einstein, instanceOf, Physicist*) could also be represented as (*Physicist, hasInstance, Albert_Einstein*). Furthermore, the structure of the knowledge base is dependent on the domains from which the facts were extracted. For example, a movie-oriented knowledge base might have a lot of facts about actors but very few facts about politicians. An in-degree-based measure of informativeness on such a knowledge base would say that Ronald Reagan is more famous for being an actor than for being a politician (assuming that the knowledge base contains facts about Reagan). In Section 4.5, we compare our scoring with the scoring of BANKS [28], which exploits the in-degree of nodes to capture their importance.

In summary, confidence and informativeness are two complementary components of our model. The *confidence* expresses how certain we are about a specific fact – independent of the query and independent of how popular the fact is on the Web. The *informativeness* captures how useful the fact is for a given query. This depends also on how visible the fact is on the Web. In this spirit, our definition of informativeness differs from the information-theoretic one, which would consider less frequent facts as more informative. The latter is captured by our background model, which will be discussed at the end of this subsection.

Our definition of informativeness depends on the query formulation. For example, the fact (*Bob_Unknown, instanceOf, Physicist*) would be less informative if the query asked for (famous) physicists (i.e., $q = (\$x, instanceOf, Physicist)$), but could be very informative if the query asked for the occupation of *Bob_Unknown* (i.e., $q = (Bob\_Unknown, instanceOf, \$x)$). Hence, our informativeness measure is asymmetric and depends on the position of the variables in the query. Therefore, symmetric information-theoretic measures, such as PMI (point-wise mutual information), would not be an adequate choice for the estimation of informativeness.

ESTIMATING COMPACTNESS The *compactness* of answers is implicitly captured by their likelihood given the query. This is because the likelihood of an answer graph is the product over the probabilities of its component facts. Therefore, the more facts in an answer graph the lower its likelihood and thus its compactness.

For example, for the query that asks for a broad connection between Margaret Thatcher and Indira Gandhi, the answer graph stating that they are both *prime-ministers*, is more compact than the answer that they are both prime-ministers of English-speaking countries.

4.3. A Framework for Ranking with Entities and Relationships

THE BACKGROUND MODEL We turn to estimating $\tilde{P}(q_i)$, which plays the role of giving different weights to different fact templates in the query. This is similar in spirit to the idf-style weights for weighting different query terms in traditional statistical LMs. For a single-term query the idf part would just be a constant shift or scaling, which does not influence the ranking. But for multi-term queries, the idf weights give more relevance to those query terms that are less frequent in the corpus.

In our model, we view a fact template from the query as a pattern from the knowledge base. Consider the fact template (Albert_Einstein, instanceOf, $x). As a pattern this template fits several facts from the knowledge base, i.e., (Albert_Einstein, instanceOf, Physicist), (Albert_Einstein, instanceOf, Cosmologist) (Albert_Einstein, instanceOf, Philosopher), etc. Intuitively, the more variables a fact template has, the more matches can be found in the underlying ER graph, and the more frequent the corresponding pattern is in the knowledge base. Hence, in analogy to traditional $tf * idf$ models, the value $\tilde{P}(q_i)$ gives more relevance to fact templates with fewer variables, or in other words, to less frequent patterns from the knowledge base.

4.3.4 Related Ranking Models

Probabilistic, LM-based ranking models have been recently used in the context of entity ranking [70, 124, 133, 142, 143]. The general idea is to view the LM of an entity e as the probability distribution of words seen in the context of e. Given a keyword query q the score of e with respect to q is determined as the probability of the LM of e having generated q.

The extension to a general method for ranking facts is not straightforward and is not addressed by the above approaches.

Libra [120] uses a statistical LM to rank structured records about authors, publications, conferences, journals, and communities. The records are ranked with respect to keyword queries. The idea is to view each record as a bag of words and compute the probability that a record generates the keywords of the query. This is very different from NAGA's graph-based querying and ranking framework.

NAGA's ranking model is a novel and promising application of statistical LMs to the setting of facts and fact templates. It opens up new perspectives for advanced ranking strategies over ER graphs. For example, [66] very recently extended NAGA's query and ranking model to support graph-based queries augmented with keywords. The assumption is that each fact f of the knowledge base is associated with a set of textual terms derived from the witness pages of f. For instance, when we are looking for a certain movie associated with the words "needle park", starring Al Pacino, we can simply pose the query (Al_Pacino, actedIn, $x){needle park}. In this case, the proposed ranking model would give a higher relevance to facts that match the query template and are related to the keywords "needle park", resulting to higher rank for the fact (Al_Pacino, actedIn, The_panic_in_needle_park). The ranking algorithm derives an LM for the query and an LM for the answer graph. Both LMs are derived from Web-based co-occurrence statistics for facts. The LM of the query graph is in addition dependent on the co-occurrence of facts with the query keywords. Finally, the answer graphs are ranked in increasing order of the Kullback-Leibler divergence (measure for the difference between two probability distributions) between their LMs and the LM of the query.

4.4 The NAGA Engine

4.4.1 Architecture

We have implemented a complete prototype system of the NAGA engine in Java. The system architecture of NAGA is depicted in Figure 9.

BACK-END The backend consists of the knowledge base, YAGO, which is organized as an ER graph of facts, stored in a database. For each fact, YAGO knows the URLs of its witnesses. The query processing component combines different algorithms, e.g., Algorithm 1, STAR (see Chapter 5), MING (see Chapter 6), to handle user queries. The subgraphs from the knowledge base that match the user query are ranked by the ranking component. The latter derives co-occurrence statistics for entity pairs (as described in Section 4.3.3) by posing queries to a keyword search engine.

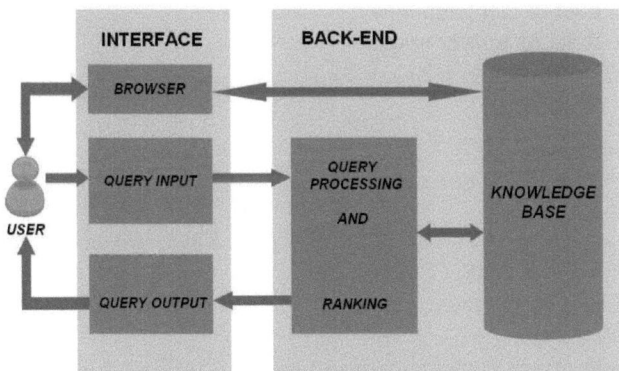

Figure 9: System Architecture

USER INTERFACE The user interface contains facilities for both the casual as well as the expert user. The expert user can use a textual input box to enter the query templates. The casual user can use the input box to enter simple queries, and can then switch to the browser. The browser renders a hyperbolic visualization of the knowledge graph. A use case for the browser could be the following. The user could start with a simple query, e.g., (Albert_Einstein, isA, $x). NAGA will return a ranked

4.4. The NAGA Engine

list of answers to this query. If the user clicks on one of the answers, an applet-based hyperbolic browser will be presented. In the beginning, the browser will contain a visualized subgraph from the knowledge base, containing the answer clicked by the user together with other facts about Einstein. The user can then successively browse the neighborhoods of the visualized entities. Additionally, by double-clicking a visualized entity the user will be shown the Wikipedia page of that entity.

4.4.2 Implementation

The facts of the knowledge base are stored in a database table with the schema *Facts*(ID, RELATION, ENTITY1, ENTITY2, CONFIDENCE). A high-level overview of NAGA's query processing algorithm is shown in Algorithm 1

ALGORITHM 1: queryResults(Q)
Input: Query $Q = (V_Q, l_{Ent}^Q, E_{Rel}^Q)$
Output: A set of answer graphs
1 normalize Q into $Q' = (V_{Q'}, l_{Ent}^{Q'}, E_{Rel}^{Q'})$
2 RETURN templateResults($Q', E_{Rel}^{Q'}$)

templateResults(Q, E)
Input: A query graph $Q = (V_Q, l_{Ent}^Q, E_{Rel}^Q)$,
a set of fact templates E
Output: A set of answer graphs
1 IF $E = \emptyset$ THEN
2   RETURN $\{Q\}$
3 END IF
4 Results = $\emptyset$
5 FOR EACH match e' of a template $e \in E$
6    $r_{e'}$ = templateResults($(V_Q, l_{Ent}^Q, E_{Rel}^Q - e + e'), E - e$)
7   IF $r_{e'} \neq \emptyset$ THEN
8   Result = Result $\cup$ $r_{e'}$
9   END IF
10 END FOR
11 RETURN Results

We first pre-process the given query into a normalized form (line 2, Function queryResults) by applying the following rewritings: first, because we allow users to use words for referring to entities, we add an additional edge labeled with *means* for each bound vertex, e.g., the query (*Einstein, hasWonPrize, $x*) becomes (*"Einstein", means, $Einstein*); (*$Einstein, hasWonPrize, $x*).

4.4. The NAGA Engine

Second, we translate the pseudo-relation *isA* into its explicit form *instanceOf subclass**, e.g., the query (*Albert_Einstein, isA, $y*) becomes (*Albert_Einstein, instanceOf subclass*, $y*). This allows the user to ask for all classes Einstein belongs to, without the need to know about regular expressions.

The main function of the query processing algorithm is templateResults. It is given a preprocessed query graph and a list of templates to be processed. Initially, the templates are edges of the query graph. We pick a template (line 6) and identify all possible matches in the knowledge base. For each possible match, we construct a refined query graph by replacing the fact template by the match (represented by the expression $E_{Rel}^{Q} - e + e'$). Note that the match e' can be a sequence of facts (see Definition 3). Then, the function is called recursively with the refined query graph. Once no more query templates need to be processed, the refined query graph constitutes a result.

We identify matches for templates as follows. In case the fact template is a simple-relation template, we translate it directly into an SQL statement. This applies to templates like (*Einstein, means, $z*), (*Albert_Einstein, $r, Ulm*), or (*$x, discovered, $z*), which can be translated into simple SELECT statements over the *Facts* table. In case the template is a regular-expression template, we first expand it into allowed sequences of simple-relation templates, which are then translated into simple SELECT statements.

REGULAR EXPRESSIONS When the edge of a template is labeled with a regular expression over relations, we construct a non-deterministic finite-state automaton (NFSA) for the regular expression. To remain efficient in the query evaluation, we require that at least one of the end nodes of the regular-expression template be bound (at evaluation time). We identify the bound node v_0 of the template. Then we try to find matches for the regular-expression template starting from v_0 (i.e., the search space of matches is explored starting from v_0). Hence, in case v_0 is not the source node of the template, but the target node, we reverse the transitions of the automaton. Consider the regular-expression template (*$x, (bornIn|livesIn|isCitizenOf) locatedIn*, Germany*). Figure 10 depicts the representation of the corresponding NFSA and its inversion.

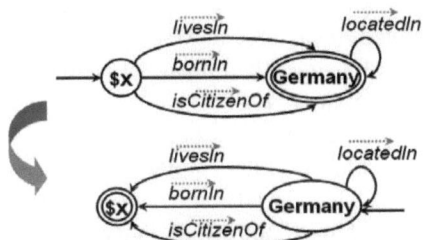

Figure 10: NFSA construction

The directions of the transitions in the NFSA should not be confused with the

4.4. The NAGA Engine

directions of the relationships in the knowledge base. In Figure 10, the direction of the relationships is depicted by the dashed arrow on top of the relationship labels. We can exploit the NFSA to expand our initial regular-expression template as follows.

Starting from state v_0 in the NFSA, we expand the tree of allowed template sequences that can be derived from the original template. Every state that can be reached via one transition from v_0 becomes a child node of v_0. Those nodes that correspond to final states in the NFSA become leaves in the tree. Then we continue the procedure successively for the children of v_0 that are no leaves. An example of such a tree is depicted in Figure 11.

Each edge in the tree is a fact template, and a path in the tree (from a leaf node to the root) represents an allowed expansion of the original template. The matches to the new template sequences are retrieved recursively as shown in Algorithm 1, starting from the upper-most templates in the tree. Typically, the regular expressions are rather simple, and we also put a limit on the expansion depth of the tree. This helps us remain efficient.

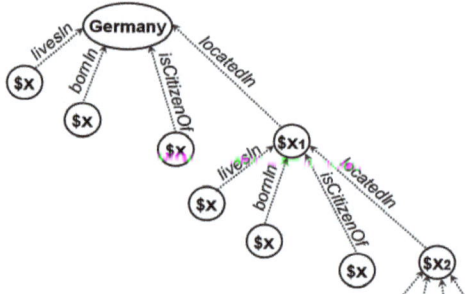

Figure 11: **Expansion of allowed sequences of templates**

4.4.3 Efficiency Aspects

Although the focus of the work presented in this chapter is not on efficiency, we have incorporated some query optimizations. First, fact templates in which the edge as well as both nodes are not labeled by a variable are processed separately, so that they do not need to be computed in each recursive call. Second, certain trivial relations (such as *smallerThan* for numbers or *before* and *after* for dates) are not stored in the database, but are computed at query time.

Queries that ask for broad connections between entities (see left-most query in Figure 7) are very challenging in terms of efficiency. As we will see in the next chapter, the underlying problem is NP hard. For this kind of queries we have developed an efficient algorithm, coined STAR, that exploits taxonomic relationships such as *instanceOf* and *subclass* in combination with local-search heuristics to retrieve the matches.

For the following evaluation, we have estimated the informativeness scores of facts in the result graphs by posing queries to a search engine (see Section 4.3.3, Equation (4.11)). Although the scores derived this way nicely capture our intuition of informativeness, it is very inefficient to do this computation for every possible answer at query time. Hence, for our online demo of NAGA [8], we have precomputed informativeness scores for facts (i.e., the needed co-occurrence statistics) from inverted indexes on the Wikipedia articles. The implementation of these scores will be explained in detail in Chapter 6, Subsection 6.2.1.

4.5 Experimental Evaluation

To evaluate NAGA's search and ranking behavior, we conducted an extensive user study that compares NAGA's performance with the performance of Google, Yahoo! Answers, and START [10, 79]. We also compared NAGA's statistics-based scoring mechanism with the one of BANKS [28], which relies on the structure of the underlying graph to derive the score of an answer graph.

4.5.1 Setup

SCORING PARAMETERS AND RANKING DESIDERATA As explained in Section 4.3 the parameters of the ranking model allow emphasizing the *confidence* or the *informativeness* of results, while at the same time, the *compactness* of answers is implicitly promoted. By means of the background model $\tilde{P}(q_i)$, the parameter α can be used to give different weights to different fact templates of a query. In a search scenario where the user is solely interested in informative matches with facts that have high confidence, α can be set to 1. For our study, we focused on the user-perceived contribution of the above desiderata to NAGA's ranking behavior. To this end, the parameter β can be used to formulate a more flexible scoring, in which either confidence or informativeness is given a higher emphasis.

For example, if we were looking for a drug that heals malaria, we would want to emphasize confidence more than informativeness, i.e., we would not be interested in famous drugs for malaria, but in drugs that have high associated confidence for healing the disease. If we wanted to find out new meanings associated with a word, we would emphasize the informativeness rather than the confidence. This would promote information that appears in possibly low-confidence sources, e.g., revealing that the word *Kleenex* (which is a trademark) is used by many people with the meaning of *tissues*.

Empirical examples for the influence of the parameter β on NAGA's ranking behavior are the following.

Consider the query ($x, isA, Physicist$). For this query, we expect answers about famous physicists at the top of the ranked list. If we choose to rank by confidence alone, i.e., by setting $\beta = 1$, we get less known physicists as the top results, while the famous ones, e.g., *Albert Einstein, Niels Bohr*, etc., are ranked lower in the list.

4.5. Experimental Evaluation

This happens because we can be equally confident that less known physicists are physicists, as we are for famous ones.

If we enable the informativeness component, by setting $\beta = 0.5$, (which gives equal weight to confidence and informativeness), the top three results are about the famous physicists *Albert Einstein, Niels Bohr* and *Max Planck*, followed by *Marie Curie* and *Blaise Pascal*. Thus our informativeness aspect plays a very important role in satisfying the information demand latent in the query.

We can observe the same effect for the query (*Albert_Einstein, isA, $x*). If we set $\beta = 1$, the top result is about Albert Einstein being a vegetarian. Setting $\beta = 0.5$, the top results are about *Albert Einstein* being a physicist, cosmologist, philosopher, etc.

For our experiments, we set β to the balanced value 0.5 giving equal weight to informativeness and confidence.

Benchmark	Question with NAGA translation
TREC	When was Shakespeare born? (*Shakespeare, bornOnDate, $x*) In what country is Luxor? (*Luxor, locatedIn*, $x*) (*$x, isA, country*)
SSearch	In which movies did a governor act? (*$y, isA, governor*) (*$y, actedIn, $z*) (*$z, isA, movie*) List movies directed by Madonna's husband? (*$x, isMarriedTo, Madonna*) (*$x, directed, $y*)
OWN	List some lakes located in Africa. (*$x, isa, lake*) (*$x, locatedIn*, Africa*) What do Albert Einstein and Niels Bohr have in common? connect(*Albert_Einstein, Niels_Bohr*)

Table 1: Sample queries

BENCHMARKS We evaluated NAGA on three sets of queries. Sample queries from each of these sets are shown in Table 1. The complete query benchmarks are given in the appendix.

- TREC 2005 and TREC 2006 provide standard benchmarks for question answering systems. Out of this set, we determined questions that can be expressed by the current set of NAGA relations. We obtained a set of 55 questions. We will denote this query set by **TREC**. Note that although NAGA knows the relations used in the questions, the knowledge graph does not necessarily have the data instances to answer them.

4.5. Experimental Evaluation

- The work on SphereSearch [77] provides a set of 50 natural language questions for the evaluation of a semantic search engine. Again, we determined 12 questions that can be expressed in NAGA relations. We will refer to this query set as **SSearch**.

- Since, to the best of our knowledge, we are the first to utilize regular expressions over general relations in a graph-based query language, we had to provide corresponding queries ourselves. We constructed 18 corresponding natural language questions. This query set will be denoted by **OWN**.

COMPETITORS Considering the fact that established search and question answering (QA) systems use different corpora, data models, query languages and rankings, the evaluation becomes very difficult. Nevertheless, in our study we try to cover a broad spectrum of retrieval systems and techniques, by comparing ourselves to state-of-the-art systems. As competitors, we chose Google (search engine), Yahoo! Answers and START (QA systems). Furthermore, in order to have a homogeneous evaluation of NAGA's scoring mechanism, we compare it to the one used by BANKS [28] – an established engine for searching over relational database graphs.

It is clear that these systems are considerably different. Google is designed to find Web pages, not to answer questions. Nevertheless, it is a robust competitor, because of its large amount of indexed Web pages. It is also tuned to answer specific types of questions (e.g., *When was Einstein born?*) directly by its built-in QA system.

Yahoo! Answers has its own corpus of questions and corresponding answers (provided by humans). Usually, the answers are also rated by users. For a given question, Yahoo! Answers first checks whether it is in the corpus of already-answered questions. If so, the answers are ranked by their ratings and returned to the user. In case the question is not present in the corpus, it is given free as an open question that can be answered by the community of users.

START is an established QA system, which employs natural-language-processing techniques to analyze and understand the meaning of natural language questions. The answers are retrieved from a background corpus containing information gathered from the Web.

BANKS performs keyword search over the graph-oriented representation of a database. The nodes of the graph represent tuples from database tables and the edges represent foreign-key relationships between tuples. The answers to a query are graphs containing the query keywords. BANKS computes the score of a result graph R as an interpolation of its overall node score $Nscore(R)$ and its overall edge score $Escore(R)$. Both values are directly inferred from the underlying graph. The value $Nscore(R)$ is computed as the average of the node scores in R, where the score of each node is proportional to its in-degree. The value $Escore(R)$ is computed as $1/(1 + \Sigma_e d(e))$, where $d(e)$ represents a distance weight between the two end nodes

4.5. Experimental Evaluation

of the edge e in R. This way, the value $Escore$ prioritizes smaller answer graphs; this is similar in spirit to our compactness criterion.

To evaluate NAGA's scoring function explicitly, we compare NAGA's scoring mechanism with the one proposed for BANKS. For this purpose, we integrated the BANKS scoring function into the NAGA engine and compared it to NAGA's own scoring mechanism. To this end, we converted the confidence values of facts in the answer graphs to distance weights (i.e., the higher the confidence the smaller the distance, and vice-versa), as they are needed as edge scores for the BANKS measure.

All the questions were posed to Google, Yahoo! Answers, START and NAGA (with its own scoring and the BANKS scoring, respectively). While for Google, Yahoo! Answers, and START the queries were posed in their original natural language form, for NAGA the queries were posed in their graph form (see Table 1). This type of comparison is influenced by several aspects. First, the results returned by a system depend on how precisely the questions can be formulated. Second, it depends on the size of the knowledge base that the system uses. Third, the comparison measures the quality of the ranking of a system. Clearly, NAGA has an advantage over Google, Yahoo! Answers, and START, because of its graph-based query language. At the same time, Google and Yahoo! Answers have a massive advantage over NAGA, because they are commercially operated systems that can search the whole Web (Google) or have a huge corpus of several millions of answered questions (Yahoo! Answers), or are explicitly designed to answer questions (START).

4.5.2 Measurements

For each question, the top-ten results of all systems were shown to human judges. On average, every result was assessed by 20 human judges – students who were not involved in this project. For each result of each system, the judges had to decide on a scale from 2 to 0, whether the result is highly relevant (2), correct but less relevant (1), or irrelevant (0).

NAGA answers queries by finding matches in the knowledge graph. For example, for a query such as (*Albert Einstein, bornOnDate, $x*), NAGA returns only the result (*Albert Einstein, bornOnDate, 1879-03-14*). Hence the direct comparison to the other systems in terms of the well known precision-at-10 ($P@10$) measure would be misleading. Therefore we chose a measure that is not dependent on the number of results returned by the system for a given query, and which can additionally exploit the rank and the weight of relevant results in the result list. This measure is the *Normalized Discounted Cumulative Gain (NDCG)*.

NDCG The NDCG measure was introduced by [90] and is intensively used in IR benchmarking (e.g., TREC). It computes the cumulative gain the user obtains by examining the retrieved results up to a fixed rank position. The NDCG rewards result lists in which highly relevant results are ranked higher than marginally relevant ones.

4.5. Experimental Evaluation

The intuition is that the lower a relevant result is ranked, the less valuable it is for the user, because the less likely it is that the user will examine the result. Thus this measure not only estimates the number of relevant results in the ranked list, but also penalizes relevant results that are ranked low in the list.

Given a query and a ranked list of results $r = r_1, \ldots, r_n$, the *gain* G_i of the result at rank i is the judgment of the user (on the scale from irrelevant (0) to highly relevant (2)). From $G_1, \ldots, G_n$, one derives the *Discounted-Cumulative-Gain vector* $\overrightarrow{DCG_r}$, which is defined recursively as follows:

$$\overrightarrow{DCG_r}[i] = \begin{cases} G[1] & \text{if } i = 1; \\ \overrightarrow{DCG_r}[i-1] + \frac{G[i]}{\log i} & \text{otherwise.} \end{cases}$$

The value $DCG_r = \overrightarrow{DCG_r}[n]$ is the *Discounted Cumulative Gain* of the ranking r. Now, one constructs the ideal ranking $r' = r'_1, \ldots, r'_n$, in which a more relevant result always precedes a less relevant one. The Discounted Cumulative Gain DCG_r is then normalized by this maximum value $DCG_{r'}$, yielding the NDCG for r:

$$NDCG_r = \frac{DCG_r}{DCG_{r'}}$$

We average the NDGC for one query over all user evaluations for that query and average these values over all queries.

PRECISION@1 We also considered the *precision at one* ($P@1$) to measure how satisfied the user was on average with the first answer of the search engine. $P@1$ is the number of times that a search engine provided a relevant result in the first position of the ranking, weighted by the relevance score (0 to 2), and normalized by the total number of evaluations multiplied with 2 (i.e., the maximum relevance score).

To be sure that our findings are statistically significant, we compute the Wilson confidence interval for the estimates of NDCG and $P@1$. We report confidence intervals for a confidence level of $\alpha = 95\%$.

4.5.3 Results and Discussion

Table 2 shows the results of our evaluation. For the TREC query set, Google performs relatively well. It has a high NDCG and in the majority of cases, the first hit in its result ranking was already a satisfactory answer. The reason for this is that the TREC questions are mostly of basic nature, i.e., factoid or list questions (see Table 1) and Google can answer a major part of them directly by its highly precise built-in question answering system. In contrast, Yahoo! Answers performs less well. Very often, it retrieves answers to questions that have only the stop-words in common with the question posed. In many cases, it does not deliver an answer at all. START

4.5. Experimental Evaluation

performs much better than Yahoo! Answers. Whenever it has the appropriate data in its knowledge base, its answers are highly satisfactory. NAGA, in contrast, is very strong on the TREC questions and outperforms all its competitors. Although most of the TREC questions translate into simple NAGA queries, NAGA does not always have the answer to a question in its knowledge graph.

Set	#Q	#A	Measure	Google	Yahoo! Answers	START	BANKS scoring	NAGA
TREC	55	1098	NDCG	75.88% ± 6.28%	26.15% ± 6.46%	75.38% ± 5.31%	87.93% ± 3.95%	**92.75%** ± 3.11%
			P@1	67.81% ± 6.87%	17.20% ± 5.52%	73.23% ± 5.46%	69.54% ± 5.63%	**84.40%** ± 4.42%
SSearch	12	343	NDCG	38.22% ± 11.22%	17.20% ± 8.54%	2.87% ± 2.87%	88.82% ± 6.80%	**91.01%** ± 6.07%
			P@1	19.38% ± 8.98%	6.15% ± 5.01%	2.87% ± 2.87%	84.28% ± 8.00%	**84.94%** ± 7.84%
OWN	18	418	NDCG	54.09% ± 11.29%	17.98% ± 8.54%	13.35% ± 6.92%	85.59% ± 6.75%	**91.33%** ± 5.28%
			P@1	27.95% ± 10.10%	6.57% ± 5.13%	13.57% ± 6.97%	76.54% ± 8.25%	**86.56%** ± 6.54%

#Q – number of questions
#A – total number of assessments for all questions

Table 2: Results

The questions from the query set SSearch are of a more sophisticated nature. They ask for non-trivial combinations of different pieces of information. Consequently, both Google and Yahoo! Answers perform on these questions worse than on the TREC questions. START performs poorly here, often because it does not understand the question (it tries to parse proper names as English words) and often because it does not know the answer. NAGA, in contrast, excels on these questions, because it makes full use of its graph-based query language.

On the queries from the set OWN, Google again performs relatively well. This is because the questions mostly ask for a broad relationship between two individuals. Google can answer these questions by retrieving Web documents that contain the two corresponding keywords. In many cases, these answers were satisfactory. Yahoo! Answers had again difficulties. START could not answer questions that ask for the broad relationship between two entities (no matter how we phrased the question) and therefore often failed. NAGA delivers good results for the majority of questions and clearly outperforms the competitors.

As shown in Table 2 (columns 8, 9) NAGA's scoring mechanism outperforms the scoring mechanism of BANKS. As already discussed, the BANKS scoring relies solely on the graph structure, which is not enough to capture informativeness. When asked for (famous) politicians, the BANKS scoring returns Albert Einstein as the first result. For the query (Albert_Einstein, isA, $x) the BANKS scoring returns person as the first result. This is because of the high in-degree of the nodes representing the entities `Albert Einstein` and `person` in the knowledge graph. NAGA, instead, captures the notion of informativeness in the overwhelming majority of the cases. It returns `Barack Obama` as the first result, when asked for famous politicians, and for the query (*Albert_Einstein*, *isA*, *$x*), the first answer is `physicist`.

Although Google and Yahoo! Answers could not capture the intended meaning of many questions from our benchmarks, they were very efficient and returned results within milliseconds. NAGA answered the majority of the queries from our benchmark in less than a second. Its runtime is comparable to that of START (although slower for regular-expression queries); but note that for each query, NAGA had to compute the scores of the answers at query time. The evaluation of query predicates with regular expressions over large ER graphs is a difficult task, especially when ranking is needed. Future research in this direction should investigate the integration of advanced indexing and top-k-processing techniques (e.g., [119, 88]) into graph-based search systems.

4.6 Conclusion

In this chapter, we presented the NAGA search engine, which shifts the retrieval focus from Web pages to knowledge. It does so by building on an expressive graph-based search framework that supports queries with entities and regular expressions over relationships. Its powerful ranking model integrates the notions of confidence, informativeness, and compactness in a principled manner. The results of the user study demonstrate that NAGA retrieves answers which are superior in quality to those returned by state-of-the-art search and question answering systems.

NAGA's LM-based ranking model could be further extended to capture a user- or context-dependent notion of informativeness. An extended model would have to consider and combine various search aspects, most importantly, the short-term history and the general search interests of the user.

In general, more advanced search and ranking models should integrate the user and the search context into their framework. Such models would have to deal with more complex ER structures resulting from n-ary relationships: e.g., user A was interested in Einstein two days ago. In terms of efficiency, they should avoid materializing large numbers of results and should exploit top-k processing whenever possible. Our work on NAGA constitutes an important step towards these challenging and exciting research directions.

5 STAR

"A hidden connection is stronger than an obvious one."

HERACLITUS OF EPHESUS

5.1 Overview

Organizing information in large ER graphs and other types of networks is abundant in modern information systems. These graphs can be used to organize relational data, Web-extracted entities, biological networks, social online communities, etc. Often, the underlying data allows the expressive annotation of nodes and edges with labels, which in turn allow the semantic interpretation of nodes as entities and edges as relations. Furthermore, edge weights can be used to reflect the strengths of semantic relations between entities. Finding close relations between two, three, or more entities is an important building block for many search, ranking, and analysis tasks. From an algorithmic point of view, this translates into computing the Steiner tree between the given nodes, a classical NP-hard problem.

In this chapter, we present a new approximation algorithm, coined STAR (**S**teiner **T**ree **A**pproximation in **R**elationship Graphs), for relatedness queries over large ER graphs. We prove that for n query entities, STAR yields an $O(\log(n))$-approximation of the optimal Steiner tree in pseudopolynomial runtime, and show that in practical cases the results returned by STAR are qualitatively comparable to, or even better than, those returned by a classical 2-approximation algorithm. We then describe an extension to our algorithm to return the *top-k* Steiner trees. Finally, we evaluate our algorithm over both main-memory as well as completely disk-resident graphs containing millions of nodes and tens of millions of edges. Our experiments show that in terms of efficiency STAR outperforms the best state-of-the-art database methods by a large margin, and also returns qualitatively better results.

5.1.1 Motivation and Problem Statement

MOTIVATION Many modern applications need to deal with graph-based knowledge representations. Such applications include business and customer networks managed in relational databases, networks over products, people, organizations, events that are automatically extracted from Web pages, metabolic and regulatory networks

5.1. Overview

in biology, social networks and social-tagging communities, knowledge bases and ontologies in RDF or ER-flavored models, and many more. Such graphs exhibit semantics-bearing labels for nodes and edges and can thus be seen as *semantic graphs*, with nodes and edges corresponding to entities and relationships, respectively, and edge weights capturing the strengths of semantic relationships between entities. Often, these graphs are too large to fit into main memory, such that the task of querying and analyzing them in an efficient way becomes non-trivial. An example of such a graph is the YAGO knowledge base [137, 138, 136], which has been constructed by systematically harvesting semi-structured elements (e.g., infoboxes, categories, lists) from Wikipedia. The resulting entities and relation instances have been integrated with the WordNet thesaurus [72] (see Section 2.2.1). Figure 12 shows an excerpt. Another well-known graph-based platform with a simpler structure is the IMDB movie database with movies, actors, producers, and other entities as nodes and the movie cast (information about directors, producers, composers, etc.) as edges.

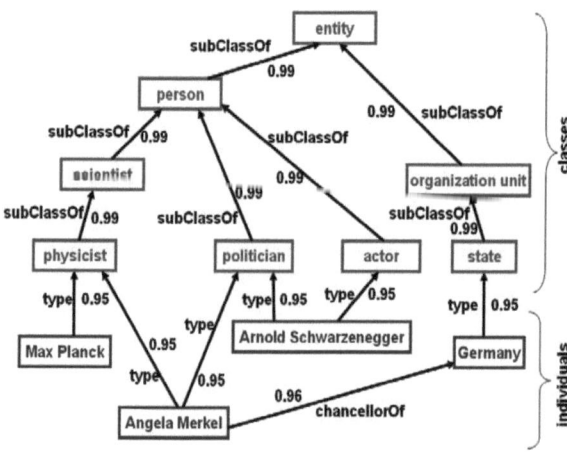

Figure 12: Example of an entity-relationship graph

Such graphs can be represented in relational or ER models, XML with XLinks, or in the form of RDF triples. Accordingly, they can be queried using languages like SQL, XQuery, or SPARQL. An important class of queries is *relatedness search*: given a set of two, three, or more entities (i.e., nodes), find their closest relations, i.e., edges or paths that connect the entities in the strongest possible way. For example, when asking *"How are Germany's chancellor Angela Merkel, the mathematician Richard Courant, Turing-Award winner Jim Gray, and the Dalai Lama related?"*, a compact (and somewhat interesting) answer is that all four have a doctoral degree from a German university (honorary doctorates in the last two cases). On movie/actor graphs, the game *"six degrees of Kevin Bacon"*[1] entails similar search

---
[1]http://en.wikipedia.org/wiki/Six_Degrees_of_Kevin_Bacon

5.1. Overview

patterns. On biological networks such as the KEGG pathway repository[2], the closest relations between the two specific enzymes and a particular gene would be of interest [110, 125, 141]. Similar queries are needed to analyze business networks between companies, their executive VPs, board members, and customers, or to discover connections in intelligence and criminalistic applications.

All the above scenarios aim at information discovery (as opposed to mere lookup), so queries should return multiple answers ranked by a meaningful criterion. Each answer can be naturally defined as a tree that is embedded in the underlying graph and connects all given input nodes. A reasonable scoring model then is some aggregation of node and edge weights over this tree. This query and ranking model has originally been proposed for schema-agnostic keyword queries over relational databases [28, 92, 20, 86]; a number of variations have appeared in the literature (see Section 5.1.2).

PROBLEM STATEMENT The formal problem that underlies these models is to compute the k *lowest-cost Steiner trees*: Given a graph $G(V, E)$, with a set of nodes V and a set of edges E, let $w : E \to \mathbb{R}_+$ denote a non-negative weight function. For a given node set $V' \subseteq V$, the task is to find the *top-k* minimum-cost subtrees of G that contain all query nodes of V', where the cost of a subtree T with nodes $V(T)$ and edges $E(T)$ is defined as $\sum_{e \in E(T)} w(e)$.

PROBLEMS WITH PREVIOUS APPROACHES Given the NP-hardness of the problem and notwithstanding the results on fixed-parameter tractability [61], as well as the tractability results on the approximate enumeration of the *top-k* approximate results [101], most prior works have resorted to heuristics, and, in fact, have typically modified the ranking model for the sake of efficiency (e.g., [82, 76, 111]). This is unsatisfying as it mixes arguments about query and ranking semantics with arguments about efficiency.

Furthermore, many of the leading database methods lack approximation or runtime guarantees (e.g., [113, 92, 28, 82]). A theoretical study conducted by the authors of [61] shows that the methods presented in [113, 92, 28, 87] turn out to have an approximation ratio of $O(n)$ where n is the number of query terms.

This work overcomes these problems by staying with the original, most natural semantics while computing near-optimal Steiner trees with practically viable runtimes. In fact, the approximation algorithm presented in this chapter even outperforms those prior methods that have worked with relaxed semantics.

5.1.2 Related Work

The problem of answering relatedness queries – queries which ask for the relations between two or more entities – has been investigated in many different applications.

---
[2]http://www.genome.ad.jp/kegg/pathway.html

5.1. Overview

Some examples are keyword proximity search over relational databases [20, 85, 86, 28, 92, 61, 82], graph search over ER, RDF and other types of knowledge bases [23, 36, 98, 139, 69], entity-relationship queries on the Web [112, 77], etc. Such applications have to deal with large graphs (sometimes with millions of nodes and edges) in general, and require not only qualitatively good solutions, but also implementations that are efficient. Our focus is on a particular kind of relatedness queries which require the system to find *top-k* connections between two or more entities. Formally, the problem of determining the closest interconnections between two, three, or more nodes in a graph is the Steiner tree problem.

The Steiner tree problem can be stated as follows. Given an edge-weighted graph $G = (V, E)$ and a set of nodes $V' \subseteq V$, called *terminals*, find a minimum-weight tree embedded in G that contains all the terminals. It has been shown that the Steiner tree problem is NP-hard. Consequently, there has been a lot of research on finding approximate solutions to this problem. The quality of an approximation algorithm is measured by the *approximation ratio*. That is, the ratio between the weight of the tree output by the algorithm and the optimal Steiner tree. The Steiner tree problem can be generalized to the Group Steiner tree problem (GST): given an edge-weighted graph $G = (V, E)$ and a set of groups $V_1, \ldots, V_k$, where each V_i contains nodes from V, find a tree in G of minimal weight such that it contains *at least* one node from each group. Obviously, an algorithm that solves the GST problem can also solve the Steiner tree problem. The GST problem can be used to model the keyword-proximity-search problem in graph structures. The assumption is that a query keyword k_i can be contained in several nodes from the underlying graph, which can be grouped to the set $V_i \subseteq V$. STAR is explicitly designed for the Steiner tree problem, as each node in a relationship graph has a unique ID (i.e., a URI) it can be addressed with.

As related work, we consider approaches to the Steiner tree and GST problem, as there exist prominent and efficient methods in both realms. In the following, we give a brief overview of related literature and compare it with our work. We do this from the perspective of the Steiner tree problem.

ALGORITHMS FOR STEINER TREE COMPUTATION Existing approaches can be categorized according to their strategies: i) distance network heuristics (DNH), ii) span and cleanup, iii) dynamic programming, iv) partition and index, and v) local search.
DNH: This heuristics [104, 118] builds a complete graph on the terminals, a so-called *distance network*. The edge weights in the distance network reflect the shortest distance between two terminals in the underlying graph. By a minimum spanning tree (MST) heuristics the distance network can be leveraged to construct a $2(1 - \frac{1}{n})$-approximation to the optimal Steiner tree. This heuristics is applicable to graphs of moderate size, which can fit into main memory. It has been emulated by other approaches for the *top-k* Group Steiner tree computation [28, 92]. The latter two approaches, however, turn out to have an approximation ratio of $O(n)$, where n is

5.1. Overview

the number of query terms (see [61]).

Span and cleanup: This heuristics [87, 129] aims at constructing the MST on the terminals by starting from an arbitrary terminal and spanning the tree stepwise until it covers all terminals. Redundant nodes are deleted in a cleanup phase. [113] exploited this heuristics by means of two different spanning strategies. In contrast to the original heuristics, each terminal is a starting point for a tree yielding a possible MST. While the first spanning strategy chooses the edge with a minimum weight to span a tree (minimum edge-based spanning), the second strategy chooses the tree the spanning of which results in a minimum cost tree (balanced MST spanning). While the method of [129] is unbound, the methods of [87, 113] turn out to have an approximation ratio of $O(n)$ (see [61]).

Dynamic programming and DPBF: The first dynamic programming approach to the Steiner tree problem was introduced by Dreyfus and Wagner [65]. It proceeds by computing optimal results for all subsets of terminals. Then the optimal result is computed for all the terminals. In [61], this heuristics is modified to a faster method, coined DPBF, for the optimal solution in the GST case. While the former work proved the fixed parameter tractability of the Steiner tree problem, the latter proved it for the GST variant. However, both methods are applicable to graphs of moderate size.

Partition and index: In this strategy, the main computation effort goes into a precomputation phase. The goal of this phase is to encounter the large size of the underlying graph by partitioning it into subgraphs (or blocks) and precomputing inter-block and intra-block shortest-path indexes. These indexes are used at query time to speed-up the query processing. Although this strategy has become quite popular in recent years [82, 111], it lacks approximation and runtime guarantees.

Local search: This heuristics has been used in the realm of the Euclidean Steiner tree problem and the parallel Steiner tree computation [29, 71]. In the first phase an interconnecting tree is built based on the distance network heuristics introduced by [104]. In the second phase the current tree is iteratively improved by considering different nodes in the underlying graph that may improve the cost of the current tree.

Our approach, STAR, cannot be fully assigned to any of the above categories. It rather combines different heuristics for efficient search-space exploration with effective local search and local pruning strategies. The main challenge here has been to provide practically viable and provable approximation and runtime guarantees. Table 3 lists the approximation ratios and runtime complexity bounds for some of the mentioned approaches with respect to the Steiner tree problem. STAR has a better approximation ratio than most of the leading database methods. In our experiments, the results produced by STAR are weight-wise comparable to the results returned by a 2-approximation or even an optimal algorithm [104, 61].

The pseudo-polynomial runtime complexity of STAR depends on the ratio between the maximum and the minimum edge weight in the underlying graph. This theoretical upper-bound boils down to a polynomial complexity bound under the

5.1. Overview

realistic assumption that the above ratio is polynomial in the size of the graph. In fact, we show in our experiments on real-life datasets that STAR outperforms some of the most efficient database methods by a large margin.

Method	Approximation ratio	Runtime complexity
BLINKS [82]	?	?
Reich & Widmayer [129]	unbounded	$O(l \cdot (m + n \log n))$
Ihler [87]	$O(l)$	$O(l \cdot n \cdot (m + n \log n))$
BANKS I [28]	$O(l)$	$O(n^2 \log n + n \cdot m)$
BANKS II [92]	$O(l)$	$O(n^2 \log n + n \cdot m)$
RIU [113]	$O(l)$	$O(l \cdot n \cdot (m + n \log n))$
Bateman et al. [26]	$O((l + \ln(n/2)) \cdot \sqrt{l})$	$O(n^2 \cdot l^2 \log l)$
Charikar et al. [39]	$O(i \cdot (i-1) \cdot l^{1/i})$	$O(n^i \cdot l^{2i})$
STAR	$O(\log(l))$	$O(\frac{w_{max}}{\epsilon \cdot w_{min}} \cdot m \cdot l \cdot (m + n \log n))$
DNH [104]	$O(2(l - \frac{1}{l}))$	$O(n^2 \cdot l)$
DPBF [61]	optimal	$O(3^l n + 2^l((l + \log n) \cdot n + m))$

n – number of nodes; m – number of edges; l – number of terminals; i – depth of tree
w_{min} – minimum edge weight in G; w_{max} – maximum edge weight in G

Table 3: Approximation ratios and runtime complexity bounds

ALGORITHMS FOR TOP-K STEINER TREE COMPUTATION Top-k Steiner tree computation has been previously studied in the context of keyword search over relational databases (see BANKS [28, 92] and BLINKS [82]).

The first BANKS paper [28] (referred to as BANKS I), addresses the GST problem on directed graphs. It emulates the DNH by running single source shortest paths iterators from each node in each group V_i, where V_i is the set of nodes which contain the keyword k_i. The iterators are expanded in a best-first strategy and follow the edges backwards. As soon as the iterators meet, a result is produced. This technique is improved in BANKS II [92] by (1) reducing the number of iterators, (2) allowing forward expansion on edges in addition to backward expansion, (3) using a *spreading-activation* heuristics which prioritizes nodes with low degrees, and edges with low weights during the expansion of iterators. However, the performance of both BANKS I and BANKS II can significantly degrade in the presence of high-degree nodes during the expansion process.

[76] makes use of the approaches of BANKS I and BANKS II to generate a first minimal-height tree that contains the query keywords. The authors show that with respect to the tree heights the *top-k* answers can be efficiently generated with provable guarantees.

DPBF [61] can be extended to a *top-k* algorithm by using the intermediate subtrees generated during the dynamic programming process to compute approximate *top-k* results.

In order to deal with graphs that may be significantly larger than main memory, the authors of [57] propose a *multigranular* graph representation that combines a condensed, memory-resident graph representation with detailed graph information

5.1. Overview

that may be cashed or stored in external memory. The goal is to minimize the IO costs during search. The authors propose different metaheuristics for retrieving the *top-k* minimum-cost Steiner trees in the multigranular graph representation. Consequently, the runtime complexity and the approximation ratio of the approach is highly dependent on the search algorithm that is plugged in the proposed metaheuristics.

Based on the notion of *r-radius Steiner graphs*, the approach of [111], EASE, exploits graph partitioning and subgraph indexing along similar lines as [82] for keyword proximity search over heterogeneous (i.e., structured, semi-structured, and unstructured) data organized as graphs. The results can be general graphs (not only trees) that contain the query keywords. The presence of a modified ranking model and subgraph indexes make theoretical implications on the runtime or approximation ratio of the approach impossible.

The recently proposed BLINKS [82] makes use of the backward search strategy of BANKS, but exploits a cost-based expansion. The authors prove that this expansion strategy, which picks the cluster with the smallest cardinality to expand next, is near-to-optimal (i.e., the number of nodes accessed by this strategy is in practice within a constant factor of the number of nodes accessed by an optimal expansion strategy). In a precomputation phase, two kinds of indexes are built to speed up the search. First, a keyword-node index is built which stores, for each keyword w, a list of nodes that can reach w along with the distance of each node from w. Second, a node-keyword index is built which stores, for each node, the set of keywords reachable from it and its distance to each keyword. However, since the proposed indexes can be too large to store and too expensive to compute, the graph is partitioned into *blocks*. The blocks are formed by partitioning the graph using node separators, also called *portals*. A high level keyword-block index is built, and more detailed indexes are built at the block level. Multiple cursors are used to perform the backward search within blocks. Whenever a portal of a block is reached, new cursors are created to explore the remaining blocks connected to this portal node.

Instead of trees, BLINKS returns $(r, \{n_i\})$ pairs, where r is the root of the result tree and n_i is a set of nodes containing the query keywords. Its scoring function differs from the usual Steiner tree scoring. It is based on the *match-distributive semantics* where the overall score of a result tree is given by the sum of the root-to-terminal paths in the tree. In general, such paths can overlap. Also in the underlying graph, there can be multiple overlapping root-to-the-terminal paths, which can be considered as candidate paths for the result tree. Figure 13 depicts such a situation. With respect to the match-distributive semantics – assuming that each edge has weight 1 – the score of the tree (represented by the bold edges) would be 9, because each root-to-terminal path contributes independently to the final score (even if paths have common edges). Given the root and the terminals, there can be different ways to construct a result tree of a certain score. But note that two different trees of the same score with respect to the match-distributive semantics can have different Steiner tree scores and vice versa. This makes the reconstruction of BLINKS trees for means of comparison with the Steiner tree semantics difficult.

5.1. Overview

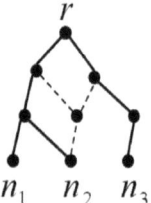

Figure 13: Example of overlapping root-to-terminal paths

Moreover, BLINKS needs to have the graph in memory to partition it and to construct the indexes, while in our approach the graph can be stored in a database and only database indexes need to be used. Finally, the performance of BLINKS is dependent on the number of portals (i.e. nodes that belong to more than one block) and the strategy for choosing them. This is because BLINKS needs to use separate cursors not just for each keyword cluster, but also for each block that it has to traverse, resulting in a high number of cursors. Hence, for a high number of portals, the performance of BLINKS suffers because of the large number of blocks that have portals in common. Although BLINKS lacks approximation and runtime guarantees, experiments show that it performs up to an order of magnitude faster than BANKS II.

5.1.3 Contributions and Outline

CONTRIBUTIONS The main contributions that will be presented in this chapter are the following.

- We present STAR, a new, efficient algorithm to the Steiner tree problem, which exploits taxonomic schema information when available to quickly produce results for l given query entities (or terminals).

- We prove that STAR has a worst-case approximation ratio of $O(\log(l))$. This improves the previously best-known approximation guarantees of $O(\sqrt{l})$ or even $O(l)$ for practically leading database methods (see [61]). In our experiments on real-life datasets, STAR achieves better results (i.e. trees of lower weight) than the ones returned by the $2(1 - \frac{1}{l})$-approximation algorithm presented in [104].

- We analyze the time complexity of the algorithm and prove that it has a pseudo-polynomial runtime (i.e., polynomial under the realistic assumption that the ratio of the maximum edge weight to the minimum edge weight is polynomial in the size of the graph.)

- We generalize STAR to an algorithm that is capable of computing approximate *top-k* relation trees for a given set of query entities.

- We compare STAR with the best state-of-the-art database methods in comprehensive main memory and on-disc experiments. STAR outperforms all opponents, often by an order of magnitude and sometimes even more.

OUTLINE The remainder of the chapter is organized as follows. In Section 5.2, we give a detailed overview of the STAR algorithm and the heuristics it uses. In Sections 5.3 and 5.4, the focus will be on the analysis of the approximation ratio and the runtime complexity of STAR. A generalization of STAR to a *top-k* approximation algorithm will be presented in Section 5.5. Finally, in Section 5.6, we present an extensive evaluation of our method. We conclude in Section 5.7.

5.2 The STAR Algorithm

As described in the introduction, we are given an undirected graph $G(V, E)$ with a set of nodes V and a set of edges E, and a non-negative weight function $w : E \to \mathbb{R}_+$, intuitively representing a distance function that is inversely proportional to the connection strength between the two end nodes of an edge. For any subgraph G' of G we denote the set of nodes of G' by $V(G')$, and the set of edges of G' by $E(G')$. Furthermore, we extend the weight function w on G' by $w(G') = \sum_{e \in E(G')} w(e)$.

Given a set $V' \subseteq V$, we are interested in finding a subgraph T of G that contains all nodes from V', such that the weight of T is minimal among all possible subgraphs of G that contain all nodes from V'. Note that inevitably, such a subgraph T has to be a tree. Furthermore, we are interested in finding the *top-k* such trees in the order of increasing weights.

Many real-world graphs come with semantic annotations such as node labels, representing entities, and edge labels, representing relations. Furthermore, these graphs may have taxonomic substructures (e.g., representing class-subclass or part-of hierarchies) indicated by the labels of the corresponding edges. The local search strategy of STAR can exploit such taxonomic backbones, when available, to efficiently find approximate solutions to the above problem. It runs in two phases. In the first phase, it tries to quickly build a first tree that interconnects all nodes from V'. In the second phase it aims to iteratively improve the current tree by scanning and pruning its neighborhood.

5.2.1 The STAR Metaheuristics

The main idea behind the STAR algorithm can be best described by a tow-phase metaheuristics. In the first phase the goal is to construct an initial tree that interconnects all terminals as quickly as possible. This can be done by:

1. Exploiting meta information about the underlying graph. In ER graphs, such meta information can be given by any subgraph that represents a conceptual

5.2. The STAR Algorithm

hierarchy (e.g., *isA* hierarchy) on the entity nodes. In general, any kind of explicit structure information about the underlying graph can be used.

2. Exploiting various heuristics for fast search space traversal.

3. Carefully precomputing and indexing interconnecting paths between some of the graph nodes.

As we will see in the next section, in its first phase, the STAR algorithm makes use of the first two strategies, to efficiently build an initial tree.

In the second phase the goal is to efficiently improve the current tree by replacing it with better solutions from its local neighborhood. This can be done by:

1. Effectively pruning the local neighborhood.

2. Exploiting heuristics for fast search space traversal.

The STAR algorithm makes use of both these strategies.

Note that ideally, one should not care about the cost of the initial tree. This would give us the freedom to use any kind of efficient heuristics for constructing the initial tree. Hence, the improvement strategy in the second phase should give us a practically viable approximation guarantee independent of the size of the initial tree. In the following we present both phases of the STAR algorithm in detail.

5.2.2 First Phase: Quick Construction of Initial Tree

In order to build a first interconnecting tree, STAR relies on a similar strategy as BANKS I [28]. But, instead of running single-source-shortest-path iterators from each node of V' (as BANKS I does), STAR runs simple breadth-first-search iterators from each terminal. The iterators are called in a round-robin manner. As soon as the iterators meet, a result is constructed. This strategy can be applied to any kind of networks, no matter whether they provide taxonomic information or not.

Unlike BANKS I, in this phase, STAR may exploit taxonomic information (when available) to quickly build a first tree, by allowing the iterators to follow only taxonomic edges, i.e., edges labeled with taxonomic relations such as *type* or *subClassOf* (see Figure 14). This way, STAR can quickly find a taxonomic ancestor of all nodes from V'. Consider the sample graph of Figure 12. Suppose that V'={**Max Planck, Arnold Schwarzenegger, Germany**}. In the first phase, STAR would construct the tree depicted in Figure 14.

5.2. The STAR Algorithm

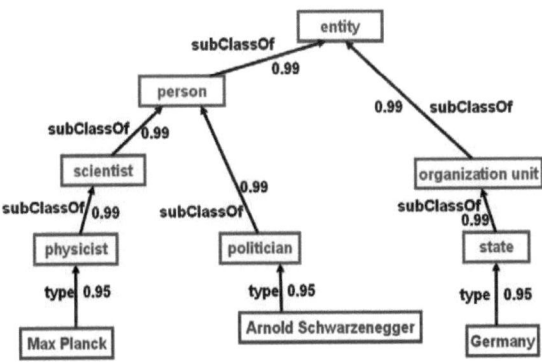

Figure 14: Taxonomic interconnection

Note that in this phase, our algorithm does not aim at minimizing the tree cost. In fact, the tree cost does not play any role in this phase.

In the following, we describe how we gradually improve the tree returned by the first phase of our algorithm.

5.2.3 Second Phase: Searching for Better Trees

In the second phase, STAR aims at improving the current tree iteratively by replacing certain paths in the tree by new paths of lower weight from the underlying graph. In the following we define which paths can be replaced.

FIXED NODES AND LOOSE PATHS Let T be a tree interconnecting all nodes of V'. We denote the degree of a node v in T by $deg(v)$. A node $v \in V'$ is called a *terminal node*, all other nodes of T are called *Steiner nodes*.

DEFINITION 5: [*Fixed node*]
A node in T is a fixed node if it is either a terminal node or a Steiner node that has degree $deg(v) \geq 3$.

Intuitively, a fixed node is a node that should not be removed from T during the improvement process.

DEFINITION 6: [*Loose path*]
A path p in T is a loose path if it has minimal length with respect to the following property: its end nodes are fixed nodes.

From the definition above, it follows immediately that every intermediate node in a loose path must be a Steiner node with degree two. Intuitively, a loose path is a path that can be replaced in T during the improvement process.

It follows immediately that a minimal Steiner tree with respect to V' is a tree in which all loose paths represent shortest paths between fixed nodes.

67

5.2. The STAR Algorithm

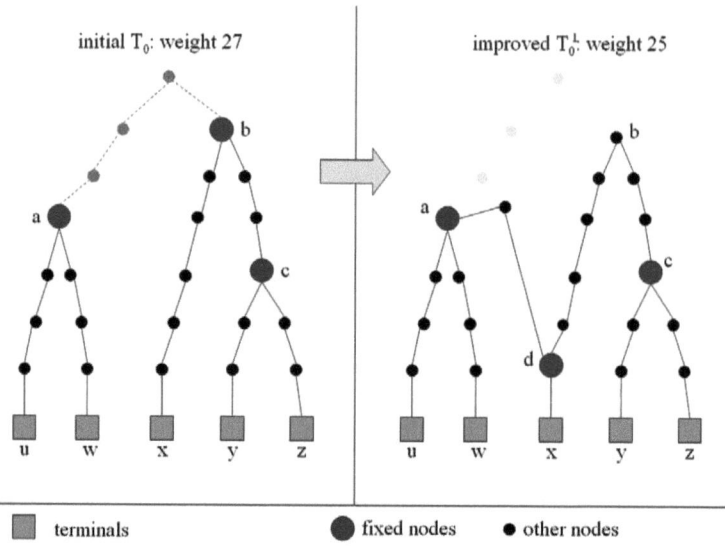

Figure 15: After first iteration

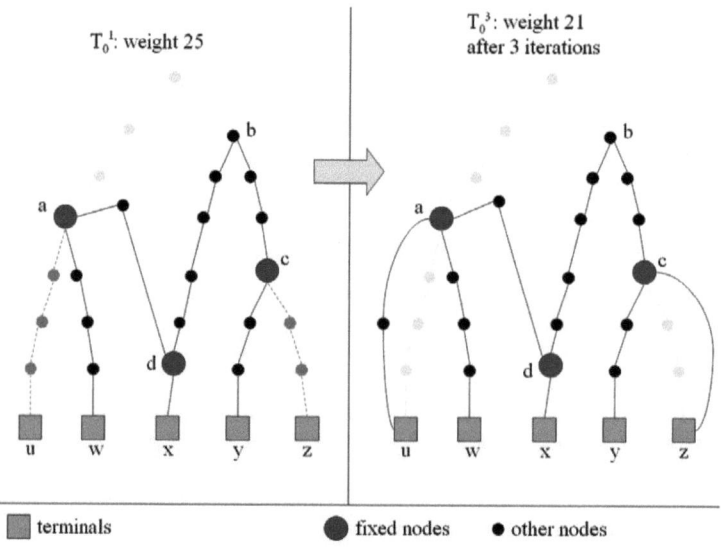

Figure 16: After third iteration

5.2. The STAR Algorithm

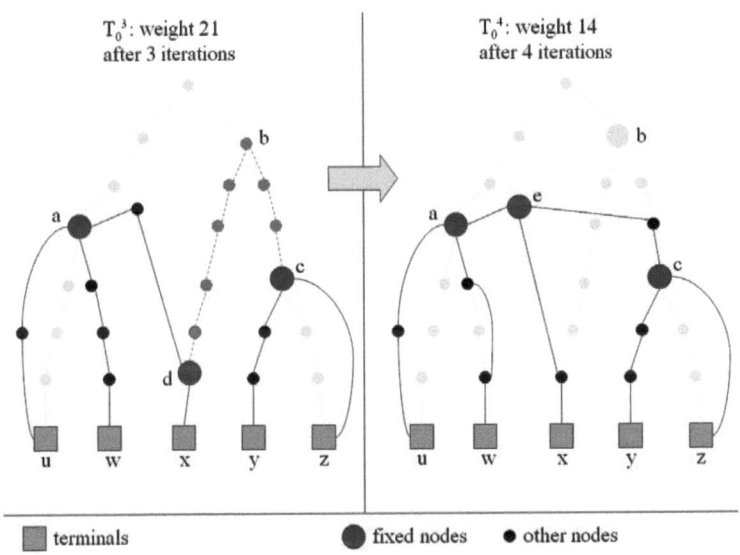

Figure 17: After fourth iteration

OBSERVATIONS In the following, for a tree T, we denote the set of its loose paths by $LP(T)$.

Removing a loose path lp from T splits T into two subtrees T_1 and T_2. In Figure 15, the removal of the loose path that connects the nodes a and b from T_0 would return two subtrees interconnecting the terminals u, w and x, y, z, respectively. Replacing a loose path lp by a new, shorter path, means computing the shortest path between any node of T_1 to any node of T_2. Note that since the end nodes of the loose path lp are fixed nodes, they are not removed when lp is removed. This means that removing a loose path that ends into a fixed node v of degree three turns v into an unfixed node, and the two remaining loose paths that had v as an end node are merged into one single loose path. In Figure 15, the removal of the loose path that connects a and b turns a and b into unfixed nodes. The loose paths that were connected to b (or to a, respectively) are merged into a single loose path. On the other hand, inserting a loose path that ends into an unfixed node v turns v into a fixed node, and the loose path that passes through v is split into two loose paths. In Figure 15, the connection of a and d by a new path turns a and d into fixed nodes. The loose path that went through d (or through a, respectively) is split into two loose paths. Hence, the number $|LP(T')|$ of loose paths in an improved tree T' is $|LP(T)| - 2 \leq |LP(T')| \leq |LP(T)| + 2$.

LEMMA 1: [*Number of loose paths in a given tree T*]
A tree T with terminal set $V', |V'| \geq 2$, has at least $|V'| - 1$ and at most $2|V'| - 3$ loose paths.

5.2. The STAR Algorithm

PROOF The proof is by induction on the number of terminals. Obviously, for a tree T with two terminals $|V'|-1 \leq |LP(T)| \leq 2|V'|-3$ holds. Let T be a tree with $|V'| > 2$. Let lp be a loose path in T. Removing lp from T splits T into two subtrees T_1 with a terminal set V'_1 and T_2 with a terminal set V'_2. By induction, our claim holds for T_1 and T_2. With the above discussion, connecting T_1 and T_2 again through lp may lead in each of the trees T_1 and T_2 to one more loose path. Hence, the overall number of loose paths in T is upperbounded by $|LP(T)| \leq |LP(T_1)| + |LP(T_2)| + 2 + 1$. On the other hand, the connection through lp may leave the number of loose paths in T_1 and T_2 unchanged, resulting in $|LP(T)| \geq |LP(T_1)| + LP|(T_2)| + 1$. Assuming that $|LP(T_1)| = 2|V'_1| - 3$ and $|LP(T_2)| = 2|V'_2| - 3$ leads to $|LP(T)| \leq (2|V'_1| - 3) + (2|V'_2| - 3) + 2 + 1 = 2|V'| - 3$. Assuming that $|LP(T_1)| = |V'_1| - 1$ and $|LP(T_2)| = |V'_2| - 1$ leads to $|LP(T)| \geq (|V'_1| - 1) + (|V'_2| - 1) + 1 = |V'| - 1$. □

FINDING AN APPROXIMATE STEINER TREE In the second phase, STAR keeps on iteratively improving the current tree T. In each iteration our algorithm removes a loose path lp from the current tree T. Consequently, in each iteration T is decomposed into two components T_1 and T_2. The new tree T is obtained by connecting T_1 and T_2 through a path that is shorter than lp (see Figures 15, 16, and 17). Hence, the inherently difficult Steiner tree problem is reduced to the problem of finding shortest paths between subsets of nodes. Heuristically, in each iteration we remove the loose path with the maximum weight in T. The reason for doing so is that we aim to effectively prune the local neighborhood of T. A high-level overview is given in Algorithm 2.

ALGORITHM 2: improveTree(T, V')
Input: Tree T produced by the first phase of STAR,
set V' of terminals
Output: Locally optimal tree
1 PriorityQueue $Q = LP(T)$ //ordered by decreasing weight
2 WHILE Q.notEmpty() DO
3    $lp = Q$.dequeue()
4   T' =replace(lp, T)
5   IF $w(T') < w(T)$ THEN
6      $T = T'$
7      $Q = LP(T)$ //ordered by decreasing weight
8   END IF
9 END WHILE
10 RETURN T

Speaking abstractly, the above algorithm greedily scans and prunes the

5.2. The STAR Algorithm

neighborhood of T for better trees. Paths that exceed the weight of the loose path upon which the current tree is being improved are pruned. Note that this method leads only to a local optimum. However, we show in Theorem 1 that this local optimum is relatively close to the global optimum.

As an example, we show how STAR would improve the taxonomic tree returned by the first phase of the algorithm (see Figure 14). In the first iteration the algorithm would remove the loose path that connects the fixed node labeled with **Germany** to the fixed node labeled with **person**. The improved tree is depicted in Figure 18. Note that since STAR aims to find closest relations between entities, it views the edges in Figures 18 and 19 as undirected.

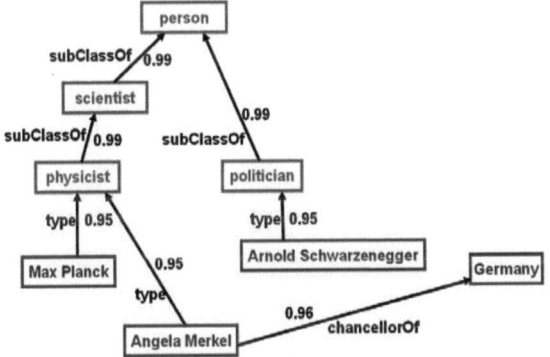

Figure 18: Result of the first iteration

In the second iteration the path connecting the fixed node labeled with **Arnold Schwarzenegger** to the fixed node labeled with **physicist** is removed. The improved tree (depicted in Figure 19) is at the same time the final tree, since no loose path can be improved. Another example is depicted in Figures 15-17.

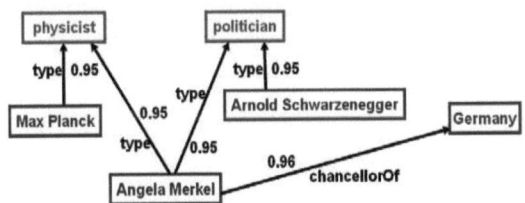

Figure 19: Result of the second iteration

The method $Replace(lp, T)$ (line 4 of Algorithm 2) removes the loose path lp from T. This removal splits T into two subtrees T_1 and T_2. Then the shortest path in G that connects any node of T_1 to any node of T_2 is determined and combined with T_1 and T_2 into a new tree T' of lower weight. For this purpose, $Replace(lp, T)$ calls another method, called $findShortestPath(V(T_1), V(T_2), lp)$, which runs one single

5.2. The STAR Algorithm

source shortest path iterator from each of the node sets $V(T_1)$ and $V(T_2)$. This method is presented in Algorithm 3. In the beginning, each of the iterators Q_1, Q_2 contains all the nodes from $V(T_1)$ and $V(T_2)$, respectively (lines 5, 6). The variables *current* and *other* (lines 7 and 8) represent the subscript indices of Q_1 and Q_2. As presented in lines 10 to 12, $Q_{current}$ points to the iterator that has minimal number of fringe nodes.

ALGORITHM 3: findShortestPath$(V(T_1), V(T_2),$ lp$)$
Input: Loosepath lp,
   subtrees T_1 and T_2 resulting from the removal of lp
Output: Shortest path from G that interconnects T_1 and T_2

```
1  FOR EACH v ∈ V(G)
2    IF v ∈ V(T₁) THEN d₁(v) = 0 ELSE d₁(v) = ∞
3    IF v ∈ V(T₂) THEN d₂(v) = 0 ELSE d₂(v) = ∞
4  END FOR
5  PriorityQueue Q₁ = V(T₁)  //ordered by incr.distance d₁
6  PriorityQueue Q₂ = V(T₂)  //ordered by incr.distance d₂
7  current = 1
8  other = 2
9  REPEAT
10   IF fringe(Q_other)<fringe(Q_current) THEN
11     swap(current, other)
12   END IF
13   v = Q_current.dequeue()
14   IF d_current(v) ≥ w(lp) THEN
15     BREAK
16   END IF
17   FOR EACH (v, v') ∈ E(G)
18     IF v' has been dequeued from Q_current THEN
19       CONTINUE
20     END IF
21     IF d_current(v') > d_current(v) + w(v, v')
22       d_current(v') = d_current(v) + w(v, v')
23       v'.predecessor_current = v
24     END IF
25     Q_current.enqueue(v')
22   END FOR
27 UNTIL Q₁ = ∅ ∨ Q₂ = ∅ ∨ v ∈ V(T_other)
28 RETURN path connecting T₁ and T₂
```

5.2. The STAR Algorithm

Intuitively, $Q_{current}$ represents the iterator that is currently expanded. This expansion heuristics is similar to the cost-balanced expansion used by BLINKS [82], which attempts to balance the number of accessed nodes (i.e., the search cost) for each iterator. It is also similar to the expansion heuristics used by BANKS II [92], which prioritizes nodes with low degrees during the expansion. However, the difference is that we consider the whole node collection in an iterator as a single node. Each iterator aims at reaching a node from the starting set (source) of the other iterator, represented by $V(T_{other})$ in line 27. Hence, in case that $Q_{current}$ points to the iterator that started from $V(T_1)$, the set $V(T_{other})$ points to $V(T_2)$ and vice versa. During the expansion, for each node v' visited by the current iterator, we maintain its current predecessor, that is, the node v from which the iterator reached v' (line 23). Again the predecessor is dependent on the current iterator. The current predecessor of v' is chosen such that the distance $d_{current}$ of v' to the source of the current iterator is minimized (lines 21-23). We maintain this distance for each visited node v' (line 22). Maintaining the predecessor of a visited node v', helps us rebuild the path from v' to the source. However, as soon as the iterator $Q_{current}$ encounters a node v that has a distance greater than or equal to the weight of the loose path lp upon which we are aiming to improve the current tree, the expansion stops (lines 14, 15). The reason for this is that all other nodes in $Q_{current}$ have a greater distance to the source than v, since the nodes in the iterators are ordered by increasing distance from the sources.

5.2.4 Discussion

A legitimate question that may arise at this point concerns the somewhat intricate framework of the STAR approach. We have presented a two-phase algorithm which exploits quite complex search strategies guided by different heuristics. Would it not be more efficient to use a much simpler search strategy that searches for an approximate Steiner tree right away?

One of the simplest search strategies in the literature is used by the BANKS I algorithm [28], which expands single-source-shortest-path iterators starting from each terminal in a best-first strategy and returns a result as soon as the iterators meet. However, this search strategy significantly degrades in the presence of high-degree nodes in the graph. As a consequence, BANKS II [92] was proposed which uses a more intricate search strategy. This time, the authors exploit a spreading-activation and a bidirectional search heuristics to avoid the explosion of the search space at high-degree nodes. In summary, it can be said that an efficient algorithm for Steiner tree search should be guided by a careful search space exploration. This, again, comes with various heuristics which need to be combined in the appropriate way. In the presence of different heuristics, the main challenge is to provide a search algorithm with practically viable approximation and runtime guarantees.

At this point, we highlight once again the main strategies of STAR, which are responsible for the efficient generation of results.

5.3. Approximation Guarantee

- **Fast construction of an initial tree.** We presented two strategies for the efficient generation of an initial tree in the first phase (see Section 5.2.2). The only goal of this phase is to build an initial tree as quickly as possible; the cost of the tree is irrelevant.

- **Effective pruning of the local neighborhood.** In Algorithm 2, we choose always the loose path with the highest weight in the current tree to remove.

- **Low cost for managing data structures.** STAR uses only two single-source-shortest-path iterators for each improvement step (see Algorithm 3); these are the only data structures used during search.

- **Smart expansion strategy for iterators.** In Algorithm 3, we use a balanced expansion strategy across iterators which prioritizes sparser regions in the search space. The balanced expansion strategy was shown to be near-to-optimal and to have a good bound on the worst-case performance [82]. Avoiding the explosion of search space at high-degree nodes was used by BANKS II [92] as an efficient search heuristics.

Despite the many heuristics it uses, STAR comes with a practically viable approximation guarantee. The next section is dedicated to STAR's approximation guarantee.

5.3 Approximation Guarantee

In this section, we prove that STAR is an $O(log(N))$-approximation algorithm, where N is the number of terminals.

Our proof has a very important implication. It entails that the approximation ratio for the cost of the final tree returned by STAR is independent of the tree constructed in the first phase.

The proof proceeds as follows. We define a mapping between each loose path in the tree returned by the algorithm, and a more expensive path in the optimal solution. Such a mapping has the property that at most $2\lceil \log N \rceil + 2$ loose paths are mapped onto a same path. Moreover, each edge in the optimal solution occurs in the range of the mapping at most twice. Hence, summing over all paths in the range of the mapping gives an upper bound (of $4\lceil \log N \rceil + 4$) on the cost of the tree yielded by the algorithm.

The process of finding such a mapping consists of two phases. First, we identify a collection of paths in the optimal tree that do not overlap too much. Then, we go back to the tree returned by the algorithm, trying not to assign too many loose paths to the same path in the optimal tree. Lemma 2 deals with this non-trivial task.

Before diving into the proof, we need some auxiliary notations. We shall denote an ordered pair by (i,j) (this means that $(i,j) \neq (j,i)$), while an unordered pair

5.3. Approximation Guarantee

will be denoted by $\{i,j\}$. For any graph G, $d_G(u,v)$ denotes the shortest distance between u and v in G. In a tree, we denote by uv the (unique) path between u and v.

Our input is an undirected graph $G = (V, E)$ and a set of terminals $V' \subseteq V$ that are to be connected. Let $N = |V'|$ (in what follows we assume $N > 2$). Let T_O be an optimal Steiner tree with respect to the set V' of terminals in the input. Let T_A be the Steiner tree returned by the STAR algorithm.

LEMMA 2: [*Mapping Loose Paths to Pairs of Terminals*]
Let $\mathcal{L}(T_A)$ be the set of *loose paths* in T_A. For any circular ordering $v_1, \ldots, v_N$ of the terminals in T_A, there is a mapping $\mu : \mathcal{L}(T_A) \to V' \times V'$ such that:

1. μ is defined for all loose paths in T_A;

2. for each loose path P with end points u and v, let T_1 and T_2 be the two trees obtained by removing from T_A all nodes in P (and their edges), except u and v; then, $\mu(P) = \{v_i, v_{i+1}\}$ for some $i = 1, \ldots, N$ and one of the nodes v_i, v_{i+1} belongs to T_1, while the other one belongs to T_2;

3. for each pair of terminals $\{v_i, v_{i+1}\}$ there are at most $2\lceil \log N \rceil + 2$ loose paths mapped to $\{v_i, v_{i+1}\}$.

PROOF For ease of presentation, we assume T_A is rooted at any arbitrary terminal node and its edges are directed from the root towards the leaves. Then, we denote by $u \to v$ a path where u is closer to the root than v. Furthermore, for any subtree T of T_A we shall denote by $\tau(T)$ the set of terminals belonging to T. The first step in defining the mapping is to find a labeling with good properties, as follows.

For each loose path $P = u \to v$ let T_u and T_v be the subtrees of T_A rooted at u and v, respectively. Let v_i and v_j be the two terminals having the minimum absolute difference $|i - j|$ among all pairs v_i, v_j, satisfying the constraints $v_i \in \tau(T_v)$ and $v_j \in \tau(T_u) \setminus \tau(T_v)$. Label P with the ordered pair (i, j). Iterate this procedure for all loose paths.

We now study some properties of this labeling. Let v_i be any terminal and let $\mathcal{P}_i$ be the path connecting the root to v_i. Consider the set of labels occurring in $\mathcal{P}_i$ of the kind (i, j), where $j > i$; let $(i, i + j_1), \ldots, (i, i + j_k)$ be the sequence of such pairs, ordered by non-decreasing j_h's. We prove that $j_{h+1} \geq 2j_h$, $h = 1, \ldots, k-1$, which together with the fact that j_h's are not larger than N implies $k \leq \lceil \log N \rceil + 1$.

Suppose by contradiction that there is h such that $j_{h+1} < 2j_h$. Consider the two loose paths labeled with $(i, i + j_h)$ and $(i, i + j_{h+1})$. Let $P = u \to v$ be the one of the two that is closest to the root.

By the definition of the labeling, $\{v_i, v_{i+j_h}, v_{i+j_{h+1}}\} \subseteq \tau(T_u)$. There are two cases, either P is labeled with $(i, i + j_h)$ or P is labeled with $(i, i + j_{h+1})$. In the former case, $v_{i+j_h} \notin \tau(T_v)$ and $j_{h+1} - j_h < j_h$. Hence, P would have been labeled with $(i+j_{h+1}, i+j_h)$. In the latter case, $v_{i+j_{h+1}} \notin \tau(T_v)$ and $j_{h+1} - j_h < j_h$, which implies

5.3. Approximation Guarantee

that P would have been labeled with $(i + j_h, i + j_{h+1})$. Therefore, in both cases we obtain a contradiction.

In other words, we just proved that in the path between the root and any terminal v_i, the number of labels of the kind (i, j), where $j > i$, is at most $\lceil \log N \rceil + 1$. From the way the labeling has been defined, as well as from the fact that there is exactly one path between the root and any terminal, it follows that in the whole tree T_A such labels can occur at most $\lceil \log N \rceil + 1$ times. Symmetrically, we can show that the number of labels of the kind (i, j) where $j < i$, is bounded by the same quantity.

In order to obtain the desired mapping the labeling is refined in the following way. Replace each label (i, j) with $(i, i+1)$ if $j > i$ and with $(i, i-1)$ otherwise. Now, drop the ordering of the pairs, that is, turn each label $(i, i+1)$ into $\{i, i+1\}$. This implies that each label can occur at most $2\lceil \log N \rceil + 2$ times. Finally, for each loose path P, define $\mu(P) = \{v_i, v_j\}$ where $\{i, j\}$ is the label of P. It is straightforward to see that the claimed three properties are satisfied. □

THEOREM 1: [*Approximation Guarantee*]
The STAR algorithm is a $(4\lceil \log N \rceil + 4)$-approximation algorithm for the Steiner tree problem.

PROOF Consider a walk on T_O that uses each edge exactly twice and that visits all nodes in T_O. Such a walk gives a circular ordering $v_1, \ldots, v_N$ of the terminals, ordered according to their first occurrence in such a walk. We have that:

$$\sum_{k=1}^{N} d_{T_O}(v_k, v_{k+1}) = 2w(T_O). \tag{5.1}$$

Using Lemma 2, we define a mapping μ with respect to the circular ordering $v_1, \ldots, v_N$. From property 2 of the mapping μ and from the termination condition of the STAR algorithm, it follows that for any loose path $P = uv$ in T_A

$$d_{T_A}(u, v) \leq d_{T_O}(\mu(uv)), \tag{5.2}$$

where $d_{T_O}(\mu(uv))$ is the distance, in the optimal solution, between the two entries of $\mu(uv)$. Finally, we can write:

$$w(T_A) = \sum_{uv \in LP(T_A)} d_{T_A}(u, v) \tag{5.3}$$

$$\leq \sum_{uv \in LP(T_A)} d_{T_O}\mu(uv) \tag{5.4}$$

$$\leq \sum_{k=1}^{N} (2\lceil \log N \rceil + 2) d_{T_O}(v_k, v_{k+1}) \tag{5.5}$$

$$\leq (4\lceil \log N \rceil + 4) w(T_O). \tag{5.6}$$

where inequality (5.4) follows from Equation (5.2), inequality (5.5) follows from property 3 of the mapping μ, and inequality (5.6) follows from Equation (5.1). ■

5.4 Time Complexity

The algorithm as it has been presented might have exponential running time. In fact, the cost of the tree might decrease at each step by an infinitesimally small amount. Fortunately, this can be solved by using a relatively simple "trick", which guarantees that at each step a significant improvement on the cost of the current tree is made.

Given $\epsilon > 0$, we introduce the *improvement-guarantee rule*, which is defined as follows. Let P be a loose path, and let P' be the path selected by the algorithm to replace P; replace P if and only if $w(P') \leq \frac{w(P)}{1+\epsilon}$. The algorithm is then iterated until no loose path can be improved.

Let $w_{\max}$ and $w_{\min}$ be the maximum and minimum cost of the edges in the input graph. The following theorem shows that the STAR algorithm with the improvement-guarantee rule is a pseudopolynomial algorithm, namely its running time is polynomial if the ratio $\frac{w_{\max}}{w_{\min}}$ is polynomial in the size of the input. Let n, m, N denote the number of vertices, edges, and terminals of the input graph, respectively.

LEMMA 3: [*Runtime with the Improvement Guarantee Rule*]
Given $\epsilon > 0$, the STAR algorithm with the improvement-guarantee rule is guaranteed to terminate in $O\left(\frac{1}{\epsilon}\frac{w_{\max}}{w_{\min}}m\right)$ steps.

PROOF Let $\bar{T}$ be the initial tree, as returned by the first phase of STAR. We have that $w(\bar{T}) \leq mw_{\max}$. At any step of our algorithm, let P be a loose path and let P' be the path selected by the algorithm to replace P. By the improvement-guarantee rule, it follows that:

$$w(P) - w(P') \geq (1+\epsilon)w(P') - w(P') \geq \epsilon w_{\min}. \tag{5.7}$$

Hence, the cost of the tree decreases at each step by at least $\epsilon w_{\min}$. This gives a bound on the number of steps k, as follows:

$$mw_{\max} - k\epsilon w_{\min} \geq 0 \Leftrightarrow k \leq \frac{1}{\epsilon}\frac{w_{\max}}{w_{\min}}m. \tag{5.8}$$

□

The next theorem shows a trade-off between the approximation guarantee of the STAR algorithm and its running time.

THEOREM 2: [*Approximation Bound with the Improvement Guarantee Rule*]
For a given $\epsilon > 0$, the STAR algorithm with the above improvement-guarantee rule is a $(1+\epsilon)(4\lceil \log N \rceil + 4)$-approximation algorithm for the Steiner tree problem. Its running time is $O(\frac{1}{\epsilon}\frac{w_{\max}}{w_{\min}}mN(n\log n + m))$.

PROOF The time-complexity bound follows from Lemma 3 and from the fact that at each step the STAR algorithm might invoke Dijkstra's algorithm at most

$(2N-3)$ times (one for each loose path, see Lemma 1). To prove the approximation ratio, it suffices to replace Equation (5.2) in Theorem 1 with:

$$d_{T_A}(u,v) \leq (1+\epsilon)d_{T_O}(\mu(uv)), \qquad (5.9)$$

and change the remaining equations accordingly. We include all steps for completeness. We have that:

$$w(T_A) = \sum_{uv \in \mathcal{L}(T_A)} d_{T_A}(u,v) \qquad (5.10)$$

$$\leq \sum_{uv \in \mathcal{L}(T_A)} (1+\epsilon)d_{T_O}\mu(uv) \qquad (5.11)$$

$$\leq \sum_{k=1}^{N}(1+\epsilon)\left(2\lceil \log N \rceil + 2\right) d_{T_O}(v_k, v_{k+1}) \qquad (5.12)$$

$$\leq (1+\epsilon)\left(4\lceil \log N \rceil + 4\right)w(T_O). \qquad (5.13)$$

∎

5.5 Approximate Top-k Interconnections

As demonstrated in Algorithm 3, the weight of the loose path lp upon which the current tree T is being improved serves as an upper bound for the weights of new interconnecting paths between the subtrees of T that result from the removal of lp from T. The final result of the STAR algorithm, as given by Algorithm 2, is a tree T in which there is no loose path upon which T can be improved.

In order to generalize STAR to an algorithm that can compute approximate *top-k* interconnections, we start from the final tree T returned by the original STAR algorithm, which is stored in a priority queue Q (see lines 1-3 of Algorithm 4). All the trees that were constructed during the improvement process of STAR are also stored in Q. They serve as possible *top-k* candidates. We then artificially relax the weight of T (line 6 of Algorithm 4) by adding a small value ϵ to its loose path weights. Such a relaxed tree can now be locally improved. Every improved tree along with all intermediate trees that led to it are inserted in the appropriate position in Q. In case the improved tree that was generated from the artificial relaxation has a greater weight than the k'th element of Q, the algorithm stops. The fact that every improvement step leads to a possible *top-k* candidate is the main efficiency ingredient in STAR's *top-k* generation strategy. We give an overview of the main steps in Algorithm 4.

5.5. Approximate Top-k Interconnections

ALGORITHM 4: getTopK (T, V', k)
Input: Tree T returned by the second phase of STAR,
set of terminals V',
parameter k representing the number of desired results
Output: Top-k approximate interconnecting trees
1 PriorityQueue Q //priority queue of trees
2 $T =$ improveTree (T, V')
3 Q.enqueue (T) //intermediate trees are already in Q
4 WHILE TRUE DO
5   $T' =$ relax (T, ϵ)
6   $T' =$ improveTree' (T', V')
7   $T =$ reweight (T')
8   IF Q.size$>= k$ AND $w(T) > Q$.get (k) THEN
9   BREAK
10   END IF
11   Q.enqueue (T)
12 END WHILE

ALGORITHM 5: relax (T, ϵ)
Input: Tree T,
relaxation parameter $\epsilon > 0$
Output: Tree T with relaxed weight
1 $T' = T$.copy ()
2 FOR EACH $lp \in LP(T')$
3   $w'(lp) = w(lp) + \epsilon$
4 END FOR
5 RETURN T'

As shown in Algorithm 5, we artificially relax the weights of each loose path lp in the current T by adding a tunable value $\epsilon > 0$. We denote the tree with the relaxed loose path weights by T'. We use these artificial loose path weights as upper bounds for the weights of new interconnecting paths between subtrees of the current tree T' that result from the removal of the corresponding loose path from T'. Then, in line 6 of Algorithm 4, we call a modification of the method *improveTree* (see Algorithm 2) on the input (T', V'). This modification takes care that during the improvement of T' upon one of its loose paths lp the new interconnecting path is not the same as lp. Note that this would always happen since the weight of lp was artificially increased, and in the underlying graph G the path lp would still be the shortest path connecting the two corresponding subtrees of T'. For this purpose, we consider only interconnecting paths that contain at least one node that is not contained in lp.

The method *reweight* (line 7) reweights the result of *improveTree'*. That is, the weight of loose paths of T' which were also loose paths in the previous tree T is set back to its original value.

The next section gives experimental evidence of STAR's quality and efficiency.

5.6 Experimental Evaluation

We compare the STAR algorithm with the most well-known algorithms for Steiner tree approximation. The algorithm [104] was the first to achieve a 2-approximation of the optimal Steiner tree. We refer to it as DNH (for "distance network heuristics"). The second algorithm is DPBF [61], a dynamic programming approach which can compute an optimal Steiner tree and performs best on a small number of terminals. The third algorithm is BLINKS [82], which is the newest and experimentally best algorithm in this field. The fourth algorithm is BANKS I [28] and its improved version BANKS II [92], which are state-of-the-art algorithms for keyword proximity search over relational data. We compared the algorithms both in terms of the quality of the returned results and in terms of their performance.

All experiments were performed on a 1.8 GHz Pentium machine with 1 GB of main memory and an Oracle Database (version 9.1) as the underlying persistent storage for all on-disk experiments. All implementations are in Java.

In this study we focus on efficiency and the goodness of Steiner trees (i.e., their weights). We do not consider the "semantic quality" or user perceived relevance of results. This aspect is orthogonal to the algorithmic focus of this work.

5.6.1 Top-1 Comparison of STAR, DNH, DPBF, and BANKS

The goal of the DNH algorithm is to compute a good approximation to the optimal Steiner tree for a given graph and given terminal nodes. The algorithm has an approximation ratio of $2(1 - \frac{1}{n})$, where n is the number of terminal nodes. STAR, by contrast, has an approximation ratio of $4\log(n) + 4$. BANKS I and BANKS II have an approximation ratio of $O(n)$. These bounds, however, are theoretical bounds for the worst case. Therefore, we studied how the above algorithms perform in practice. To compare to optimal tree weights, we also ran DPBF. To have comparable runtimes we reimplemented DPBF in Java[3].

DATASETS We use subsets of DBLP[4] and IMDB[5] for our experiments. DBLP

---

[3] The original C++ code was kindly provided to us by the authors of [61].
[4] Data downloadable from http://dblp.uni-trier.de/xml
[5] http://www.imdb.com/

5.6. Experimental Evaluation

and IMDB can be viewed as graphs in which nodes represent entities (like *author, publication, conference, actor, movie, year*, etc.), and edges represent relations (like *cited_by, author_of, acted_in*, etc.). Since the DNH and the DPBF algorithms are designed to deal with graphs that can be completely loaded into main memory, we extracted from DBLP a subgraph with 15,000 nodes and 150,000 edges (dataset DBLP).

As the qualitative performance of the algorithms can be influenced by different graph topologies, a second graph consisting of 30,000 nodes and 80,000 edges was extracted from IMDB (dataset IMDB). Since the original DBLP and IMDB do not provide any edge weights, we used random weights between 0 and 1 for both graphs. Note that since these datasets do not have any kind of taxonomic backbone, STAR uses its breadth-first heuristics for the initialization phase.

QUERIES We constructed three query sets with 3, 5 and 7 terminals, respectively. Each query set consists of 60 queries with the same number of terminals. The terminals were chosen randomly from the graph.

METRICS We compare the weight of the *top-1* tree returned by STAR (without taxonomic information) with the weight of the tree returned by DNH, BANKS I, and BANKS II on the basis of optimal scores returned by DPBF. We also measured the running times of all algorithms.

RESULTS Table 4 shows the results of our experiments on DBLP. The best values across the competitors are in boldface. Column 3 shows the average weight of the result over the 60 queries in the query sets returned by each algorithm. The average weight of the tree returned by the STAR algorithm is consistently below the average weight of the tree returned by DNH (for the same number of terminals) and also better than the scores returned by BANKS I and BANKS II. We validated the statistical significance of the superiority of STAR using a *t-test* at level $\alpha = 0.05$. In particular, STAR returns better results than DNH for this practical case, even though DNH has a better approximation ratio. Column 4 shows the average runtime of the algorithms in milliseconds. STAR determines the *top-1* tree much faster than all its competitors. The dynamic programming approach of DPBF and the spreading activation heuristics of BANKS II seem to be less adequate for the topology of the DBLP subgraph. The question marks in row 13 of the table reflect the fact that DPBF did not return a single result within 30 minutes. Table 5 shows that BANKS II significantly improves its performance relatively to its competitors on the IMDB subgraph, but is still outperformed by STAR.

Table 5 shows that for the IMDB subgraph, the scores of STAR and DNH lie very close to each other. We hypothesize that the higher edge-to-node ratio of the DBLP subgraph allows STAR to return clearly better scores than DNH on the DBLP subgraph. In a denser graph STAR has more possibilities to improve the current tree.

5.6. Experimental Evaluation

Method	# terminals	avg. weight	avg. runtime (ms)
STAR	3	0.61	**604.2**
DNH		0.7	5402.9
DPBF		**0.58**	33096.7
BANKS I		1.22	2096.3
BANKS II		1.81	3214.1
STAR	5	0.86	**960.2**
DNH		0.98	9166.7
DPBF		**0.81**	432361.5
BANKS I		1.87	3617.3
BANKS II		2.46	5797.5
STAR	7	**1.12**	**1579.6**
DNH		1.22	17430.9
DPBF		?	?
BANKS I		2.37	5945.5
BANKS II		3.42	9435.5

Table 4: Top-1 tree comparison on DBLP

Method	# terminals	avg. weight	avg. runtime (ms)
STAR	3	3.42	**1044.5**
DNH		3.37	9110.1
DPBF		**2.93**	18014.7
BANKS I		3.85	7153.4
BANKS II		5.31	4153.2
STAR	5	4.35	**1353.5**
DNH		4.33	12912.7
DPBF		**4.14**	121863.3
BANKS I		5.52	9671.4
BANKS II		7.17	5429.1
STAR	7	**5.31**	**1732.9**
DNH		**5.31**	18317.3
DPBF		?	?
BANKS I		7.47	11681.8
BANKS II		9.12	6953.7

Table 5: Top-1 tree comparison on IMDB

5.6.2 Top-k comparison of STAR, BANKS, and BLINKS

Unlike the DNH algorithm, BANKS I, BANKS II and BLINKS can compute the *top-k* results for a query – like the STAR algorithm. In this comparison we analyze the *top-k* performance of BANKS I, BANKS II, BLINKS, and STAR. We used a Java implementation of BLINKS that was kindly provided to us by the authors. BLINKS uses indexes in order to speed up the query processing time. However, in order to build these indexes and to subsequently use them during runtime, BLINKS requires the entire graph in main memory. For this reason, we used again the DBLP and

5.6. Experimental Evaluation

IMDB dataset for the comparison. As for the partitioning strategy of BLINKS, we experimented with different block sizes and chose a block size of 100 nodes for DBLP and a block size of 5 nodes for IMDB, since these block sizes gave the best results.

METRICS Since BLINKS uses a different weight metric (the *match-distributive* semantics) and returns only the root nodes of the output trees, we could not compare STAR and BLINKS by the weight of the output trees. Hence, our comparison with BLINKS is only with respect to the runtime. For BANKS I, BANKS II and STAR we also report the average scores of the output trees.

QUERIES We compared the algorithms for $k = 10, k = 50$ and $k = 100$ on the same Steiner tree problem instances. For the comparison, we constructed for each dataset (DBLP and IMDB) 60 random queries with five terminals each.

Method	top-k	avg. weight	avg. runtime (ms)
STAR	top 10	**1.57**	**1206.3**
BANKS I		2.43	5851.8
BANKS II		3.78	7895.9
BLINKS		n/a	19051.4
STAR	top 50	**2.23**	**3118.3**
BANKS I		3.12	7335.1
BANKS II		5.31	8928.3
BLINKS		n/a	21837.9
STAR	top 100	**3.01**	**4705.1**
BANKS I		4.15	9640.8
BANKS II		6.81	11071.3
BLINKS		n/a	24632.3

Table 6: **Top-k tree comparison on DBLP**

Method	top-k	avg. weight	avg. runtime (ms)
STAR	top 10	**5.21**	**1587.2**
BANKS I		6.13	10611.3
BANKS II		8.25	6619.4
BLINKS		n/a	2848.97
STAR	top 50	**6.32**	**1936.8**
BANKS I		7.21	12049.3
BANKS II		10.04	7892.2
BLINKS		n/a	3708.6
STAR	top 100	**8.07**	**2503.2**
BANKS I		9.92	13694.1
BANKS II		14.98	8873.3
BLINKS		n/a	4917.7

Table 7: **Top-k tree comparison on IMDB**

RESULTS We computed the average runtime and the average score for the retrieved *top-10*, *top-50* and *top-100* results. Table 6 and Table 7 present the runtime performance of STAR, BANKS I, BANKS II and BLINKS on the DBLP and IMDB datasets, respectively. Note that in this comparison we have discounted the times needed by BLINKS to construct the indexes. The results show that STAR outperforms its competitors in all cases. It is interesting to see that BANKS II and BLINKS perform better on the sparser IMDB graph. During search, BLINKS has to cope with a large number of cursors resulting from a large number of partitions. Whenever BLINKS reaches a portal p which belongs to multiple partitions, it has to construct a new cursor for each partition in which p is a portal. In dense datasets, it is likely that a large number of cursors are required to complete the query processing. The overhead of maintaining these cursors adversely affects the overall performance. An indication for this is given by the worse runtime performance of BLINKS on the DBLP dataset.

In contrast, STAR has to maintain only two iterators per improvement step. Furthermore, these iterators do not visit nodes that have a distance from the source that is higher than the upper bound given by the loose path to be replaced. The combination of tight upper bounds to prune the exploration with low overhead in iterators allows STAR to outperform BLINKS by a large margin.

5.6.3 Comparison of STAR and BANKS

Unlike DNH and BLINKS, BANKS and STAR can be directly applied to graphs that do not fit into main memory. Since these kinds of scenarios are realistic for the Steiner tree problem, we decided to simulate such a scenario by using a disk-resident dataset for the comparison of BANKS and STAR.

DATASET We chose the graph of the YAGO knowledge base [138]. It contains 1.7 million nodes and 14 million edges. Each edge corresponds to a fact in YAGO, and has a confidence score between 0 and 1 associated with it. We converted these confidence scores into distance measures. We store the graph in a relational database with the simple schema

$$EDGE(source, target, weight).$$

YAGO contains a DAG-shaped taxonomy of *type* and *subClassOf* edges (see Figure 12), which is exploited by STAR in its first phase to construct the initial tree.

We implemented both BANKS I [28] and its improved version BANKS II [92] in Java following their descriptions for main-memory procedures. Whenever the algorithms explore a new edge, we loaded the edge from the database. This way, BANKS and STAR were treated uniformly as far as the overhead for database calls is concerned.

QUERIES We generated 2 sets of queries with 3 and 6 terminals each. Each query set consisted of 30 queries with randomly chosen terminal nodes. We measured the performance of the algorithms for the *top-1*, *top-3* and *top-6* results.

METRICS We measured both the quality of the output trees and the efficiency of the algorithms. As for the quality of the trees, we report the *average weight* of the *top-k* results. As for efficiency, we report the running times and also the number of edges accessed during the query executions. There were several cases for which BANKS I and BANKS II did not return a result within 30 minutes and we had to stop the process. To be fair, we excluded these cases from our evaluation.

	3 terminals			6 terminals		
top-1	STAR	BANKS I	BANKS II	STAR	BANKS I	BANKS II
avg. score	**0.22**	0.260	0.234	**0.337**	0.385	0.368
avg. # acc. edges	**6981**	84171	81462	**9559**	372634	365004
avg. run time (ms)	**12440.6**	131313.6	104148.5	**15733.1**	391601.0	385401.5
top-3	STAR	BANKS I	BANKS II	STAR	BANKS I	BANKS II
avg. score	**0.428**	0.488	0.454	**1.085**	1.193	1.255
avg. #acc. edges	**18027**	153078	132141	**27085**	460521	409414
Avg. run time (ms)	**34814.7**	190547.7	156535.3	**41187.3**	483328.4	427276.3
top-6	STAR	BANKS I	BANKS II	STAR	BANKS I	BANKS II
avg. score	**2.102**	2.453	2.441	**3.315**	4.148	4.031
avg. # acc. edges	**43474**	159130	175045	**76259**	503054	491786
avg. run time (ms)	**71058.2**	197543.7	205359.6	**91157.2**	511811.0	491785.5

Table 8: YAGO: Quality of results and efficiency of STAR and BANKS

RESULTS Table 8 shows the results for the performance of STAR, BANKS I, and BANKS II. Concerning the quality of the output trees, STAR returns better results across all values for k and all sets of queries.

As for the efficiency of the algorithms, we note that STAR is an order of magnitude faster than BANKS. This is also reflected directly in the number of edges accessed by each algorithm: STAR accesses an order of magnitude fewer edges than its competitors. This clearly shows the enormous gains that can be made by exploiting the taxonomic structure of the tree to construct the initial result.

5.6.4 Summary of Results

We compared STAR to different state-of-the-art algorithms. Some of these algorithms come with specific constraints: The DNH algorithm, for example can only handle graphs that fit into main memory and can produce only *top-1* results. BLINKS uses indexes and a different metric and hence cannot give an approximation guarantee. To be fair, it should be emphasized that some of these methods were designed with broader goals beyond Steiner-tree-like relationship queries. Our comparison focuses on Steiner tree computation and is fair by giving all methods the same inputs, operating conditions and resources. In all experiments, STAR outperforms its competitors.

The reason for the efficient performance of STAR is three-fold: i) STAR uses the taxonomic structure of the graph when possible to quickly return an initial result which is then improved, ii) since STAR requires only two iterators per improvement step (independent of the number of terminals), the cost for managing data structures remains low throughout the search and generation process, and iii) STAR builds on efficient search space exploration strategies and effective search space pruning heuristics.

5.7 Conclusion

This work has addressed the problem of efficiently answering relatedness queries over entity-relation-style data graphs. The STAR algorithm can exploit taxonomic structures that are inherent in many knowledge-base graphs (e.g., the isA hierarchy) for fast computation of an initial seed solution. However, it does not depend on this option, and can use other initializations as well. Its main power for efficiency and result quality comes from a careful design that combines various search space exploration and pruning heuristics with elaborate graph-theoretical analysis.

We proved that STAR achieves an $O(\log(n))$ approximation for the optimal Steiner tree, which is significantly better than the worst-case approximation quality given by prior database methods [28, 92]. While the DNH method for in-memory graphs has a much better worst-case approximation guarantee than STAR, our experiments give evidence that STAR achieves at least the same result quality (Steiner tree weight) as DNH and other database methods or better on practically relevant datasets.

The motivation for this database-algorithmic work has been to support graph-based information retrieval and knowledge queries over large datasets in the spirit of NAGA [98], where STAR closes a big efficiency-oriented gap. STAR has been implemented as a query answering component of the NAGA system.

An interesting research direction would be the extension of STAR with partition-and-indexing strategies in the realm of BLINKS. The extension of STAR to a GST version for complex search patterns over richly annotated relationship-graphs is an equally interesting research direction.

6 MING

"All men by nature desire knowledge."

ARISTOTLE

6.1 Overview

Many modern applications are faced with the task of knowledge discovery in large ER graphs, such as domain-specific knowledge bases or social networks. An important building block of many knowledge discovery tasks is that of finding "close" relations between $k \geq 2$ given entities. We investigated this kind of knowledge discovery task in the previous chapter. A more general knowledge discovery scenario on ER graphs is that of mining an "informative" subgraph for $k(\geq 2)$ given entities of interest (i.e., query entities). Intuitively, this would be a subgraph that can explain the relations between the k given query entities. This knowledge discovery scenario is more general than the one of the previous chapter in that its focus is on whole subgraphs (and not only on trees). We are interested in measures that capture the human intuition of an informative ER subgraph. An adequate measure should favor insightful and salient relationships between the query entities.

In this chapter, we addresses this problem of mining informative ER subgraphs. We define a framework for computing a new notion of informativeness of nodes. This is used for defining the informativeness of entire ER subgraphs. We present MING (**M**ining **In**formative **G**raphs), a principled and efficient method for extracting an informative subgraph for $k(\geq 2)$ given query entities. The viability of our approach is demonstrated through experiments on real-life datasets, with comparisons to prior work.

6.1.1 Motivation and Problem Statement

MOTIVATION ER graphs are abundant in the field of knowledge representation. They come in different flavors and formats (i.e. represented through relational models, XML with XLinks, or RDF triples) and cover various knowledge domains. Examples of ER graphs are GeneOntology [5] or UMLS [14] (in the biomedical domain), SUMO [121], OpenCyc [56], WordNet [16, 72], YAGO [137, 138, 136]

6.1. Overview

(in the domain of general purpose knowledge bases), the ER graphs represented by IMDB (in the domain of movies and actors), DBLP (in the domain of Computer Science publications), and LOD [30] (for publishing interlinked Web data sets as RDF graphs), and many more.

Applications exploiting ER graphs are often faced with knowledge discovery tasks. Frequent scenarios here are those that aim to find meaningful relations between $k(\geq 2)$ entities of interest. From a graph-theoretic point of view, the goal in such scenarios would be to determine a subgraph that can explain the relations between the k entities of interest. We will interchangeably refer to these entities as *query nodes* or *query entities*. A related knowledge discovery task, namely that of finding "close" connections between $k(\geq 2)$ query entities, was investigated in the previous chapter. In contrast to the previous chapter, where the focus was on subtrees that closely interconnect the given query entities, the task considered in this chapter aims at finding whole subgraphs that capture insightful relations between $k(\geq 2)$ query entities. Corresponding queries could ask for the relations between k given biomedical entities, the connections between k criminals, the most relevant data shared by k Web 2.0 users, etc. For large ER graphs, these queries become challenging from an algorithmic as well as from a semantics viewpoint. The answer graphs should be computed efficiently, and they should be insightful by exhibiting salient facts. This challenge calls on one hand for adequate measures for capturing the semantic relatedness between the query entities, and for robust and efficient solutions on the other hand.

PROBLEM STATEMENT Formally, the general problem that motivates this chapter can be stated as follows: given a set $Q = \{q_1, ..., q_k\}, k \geq 2$, of nodes of interest (i.e., query nodes) from an ER graph G and an integer $b > k$ (representing a node budget), find a connected subgraph S of G with at most b nodes that contains all query nodes and maximizes an "informativeness" function $g_{info}(S, Q)$. Intuitively, for the given node budget b, this would be the subgraph that best explains the relations between the entities represented by the query nodes, in other words, this would be the most informative subgraph. The above problem comes with two subproblems:

1. What is a good measure for capturing the informativeness of relations between entities in ER graphs?

2. How to determine the most informative subgraph for the given query nodes efficiently?

Consider an ER graph about prominent persons with rich information about their careers, nationalities, interests, their birth and death dates, their prizes, etc. Note that the YAGO knowledge base is an example of an ER graph with such information about prominent persons. Figure 1 (in Chapter 2) depicts an excerpt from the YAGO graph. Consider the query that asks for the relation between *Max_Planck*, *Albert_Einstein*, and *Niels_Bohr*. An informative subgraph that captures their

6.1. Overview

relatedness should reveal that all three of them are physicists, scientists, Nobel Prize winners, etc, and should discourage long or obscure connections (e.g. connections through persons with same nationalities or same birth or death places as some of the query entities). Figure 20 depicts a possible answer.

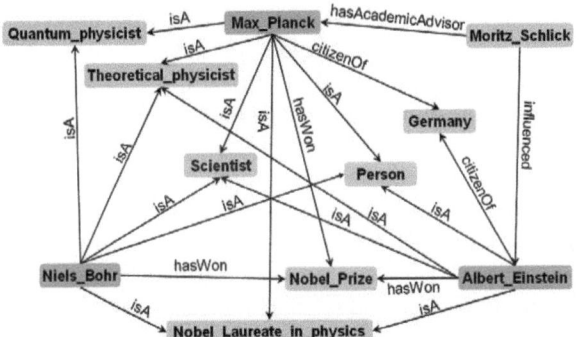

Figure 20: Answer graph returned by MING on YAGO

PROBLEMS WITH PREVIOUS APPROCAHES In previous approaches [38, 69, 73, 106, 123, 139], the notion of subgraph importance is mainly based on structural properties of the underlying graph (e.g. indegree or outdegree of a node, density or edge connectivity[1] of a subgraph, etc.). More related to our approach are techniques based on influence propagation like [69] or [139]. The approach of [69] exploits a current-flow-based algorithm and comes with an efficient two-phase solution for dealing with disk-resident graphs, but it is restricted to two query nodes. The approach of [139], CEPS, can handle more than two query nodes, and gives a random-walk-based solution for retrieving the most "central" nodes, so called *centerpieces*[2], with respect to the query nodes, but cannot be applied to disk-resident graphs in a straight-forward manner. In addition, all mentioned approaches leave aside the problem of deriving measures for capturing the semantic importance of nodes and edges in ER graphs.

Other, Steiner-tree-based, approaches [20, 28, 61, 82, 85, 86, 92, 95] have addressed the problem of retrieving the top-k minimum-cost subtrees that closely interconnect the given query nodes. Their result paradigm is tree-based. Hence, these approaches are not directly applicable to our problem of retrieving informative subgraphs. The top-k result trees can be combined into a single subgraph that interconnects the query nodes, but again, the underlying cost models are rather driven by structural properties than by the semantic importance of nodes and edges. In fact, the cost models are often modified for the sake of efficiency (see for example [82]).

In contrast, our approach gives an efficient solution for large, disk-resident ER

---
[1]Size of the minimum cut in a graph.
[2]Term introduced in [139] to describe intermediate nodes that are closely connected to most of the query nodes.

6.1. Overview

graphs, while making the semantic aspect of entities and relationships in ER graphs a key ingredient for the measure of informativeness.

6.1.2 Related Work

There are various approaches which aim at identifying important subgraphs by applying structural analysis, e.g., by identifying strongly connected, dense or frequent subgraphs [73, 74, 75, 106, 107, 123], by emulating random walks, electrical circuits or other influence propagation techniques [38, 69, 100, 139], by applying graph clustering and partitioning [32, 60, 64, 151], by computing Steiner trees [20, 28, 61, 82, 85, 86, 92, 95], etc. To our surprise, the goodness measures for subgraphs in all these approaches are guided by two main aspects: frequency of subgraph patterns, or structural properties of subgraphs. However, for ER graphs, this is not sufficient, since (1) these graphs are usually free of redundancy, which attenuates the frequency aspect, and (2) they represent only a "biased" subset of the real world, which attenuates the structural aspect. For example, an RDF database may contain a lot of facts about a special entity X just because these facts were easy to extract. This does not mean that X is in general more important than entities for which there are fewer facts in the database. This has also been observed by Ramakrishnan et. al [128] who introduce a goodness measure that goes beyond the mere structure- or frequency-based importance. However, they too, infer this new measure directly from the ER graph. We strongly believe that a goodness measure for ER subgraphs should exploit the information redundancy of the domain from which the ER graph was derived. In the following, we discuss some related approaches by focusing on the main characteristics of their goodness measures.

STEINER TREE DETECTION In contrast to the general graph-based result paradigm of the work presented in this chapter, the result paradigm in this area is tree-based. The goal is to find subtrees of the underlying graph that closely interconnect the given query nodes. BANKS I [28] and BANKS II [92] use single-source-shortest-path iterators which start from the query nodes and follow the directed edges of the graph backwards (BANKS I), or backwards and forwards (BANKS II). A result tree is produced as soon as the iterators meet. The goodness measures for their result trees are based on indegrees and outdegrees of nodes as well as on edge weights. BLINKS [82] retrieves result trees efficiently by means of subgraph partitioning and indexing. It builds on the BANKS heuristics and uses a cost model that allows the combination of subresults that were computed on different partitions. Finally, the goodness measures of STAR [95] and DPBF [61] merely build on edge weights. While STAR uses a local search strategy in combination with different search space exploration heuristics, DPBF exploits a dynamic programming strategy.

COMMUNITY DETECTION In most of the community detection approaches, the goodness measures for subgraphs build on structural properties. Gibson et

6.1. Overview

al. [73] address the emergence of communities in the Web graph. They exploit the HITS algorithm [103, 102] to determine the *top-k* hubs and authorities for a given topic. Usually, these hubs and authorities form a structurally dense and topic-specific core. Kumar et al. [106] exploit the hypothesized correspondence between communities and dense bipartite subgraphs to detect communities. Their algorithm is a two-step process – a careful enumeration and removal of small-sized bipartite cliques, followed by an apriori-style enumeration algorithm on the residual, hopefully smaller, graph. [74] presents a recursive shingle-based algorithm[3] which seeks clusters of similar Web pages that tend to link to the same destinations. Apart from detecting patterns of dense subgraphs, the algorithm can also recursively detect similarities between such subgraphs.

GRAPH CLUSTERING AND PARTITIONING [151] exploits edge connectivity to mine *closed subgraphs*[4] in a set of ER graphs. Efficient methods for identifying corresponding patterns are presented. SkyGraph [123] addresses the problem of discovering the most important ER subgraphs, where the importance of a graph is determined by its order (i.e., the number of nodes) and its edge connectivity. SkyGraph uses successive applications of the Min-Cut algorithm [81] starting with the original graph and proceeding with all produced subgraphs. Finally, a notion of subgraph domination, introduced by the authors, leads to the most important ER subgraphs.

INFLUENCE PROPAGATION More related to our approach are techniques that build on influence propagation. For a given ER graph, HubRank [38] precomputes and indexes random walk fingerprints for a small fraction of nodes, carefully chosen using query log statistics. At query time, the nodes with indexed fingerprints are exploited to compute approximate personalized PageRank vectors for a query relevant subgraph. In [69], Faloutsos et al. present an approach that emulates electrical circuits to retrieve a subgraph that captures important relations between two given entity nodes. The approach proceeds by determining a connected candidate subgraph C that contains many important connections between the two query nodes. By applying +1 voltage on one query node, the method determines (based on a current-flow measure) the subgraph S of C that contains the most important interconnections between the two query nodes. The approach is generalized in [139] by a method coined CEPS, which can be applied to any number of query nodes. The problem addressed there is that of finding *centerpieces*, i.e., intermediate nodes that are closely connected to most of the nodes from a node set Q of query nodes. Based on random walks with restarts from each of the query nodes, the k most central nodes with respect to Q are retrieved. The method is extended to extract a connected subgraph, which, as reported, captures the intuition about important relations between the nodes of Q. However, CEPS is not applicable to disk-resident

---
[3]Text mining method for estimating the similarity between Web pages by examining their feature overlap.
[4]A graph is closed if and only if there is no supergraph that has the same support (i.e., frequency).

6.2. ER-Based Informativeness

graphs in a straight-forward way.

In [128] the authors address the same problem as [69]. A current-flow-based algorithm for subgraph generation is combined with different heuristics for capturing the specificity and the selectivity of relations and entities (e.g., the entity *Theoretical_Physicist* is more specific than *Physicist*, accordingly a fact of the form (*Person, livesIn, City*) is less selective than a fact of the form (*Person, isMayorOf, City*)). However, all measures behind these heuristics are directly inferred from the graph at hand. We argue that in practice, this is not sufficient, since ER graphs represent only a limited fraction (usually restricted to certain domains) of the real world.

6.1.3 Contributions and Outline

This chapter addresses the problem of finding a subgraph that can explain the relations between $k(\geq 2)$ query nodes from an ER graph. We compute the most informative subgraph in a two-phase approach, coined MING (**M**ining **In**formative **G**raphs), that can efficiently deal with disk-resident ER graphs. In its first phase MING extracts a connected candidate subgraph that contains many important connections between the query nodes. In the second phase MING uses a random-walk-based learning method to determine the most informative answer graph. Our main contributions are the following:

- We give a clean notion of informativeness for nodes in ER graphs. Our informativeness measure builds on a natural extension of the random surfer model that underlies PageRank [33]. This measure is exploited to capture the informativeness of entire ER subgraphs.

- We present MING, a robust and efficient method for mining and extracting most informative subgraphs that best capture the relations between $k(\geq 2)$ query entities.

- We demonstrate the viability of our approach in an extensive evaluation on real-life datasets, based on user assessments and in comparison with state-of-the-art extraction techniques for ER graphs.

The remainder of the chapter is organized as follows. Section 6.2 introduces the notion of informativeness for ER graphs. Section 6.3 is dedicated to our subgraph mining and extraction algorithms. We present the experimental evaluation of our approach in Section 6.4, and conclude in Section 6.5.

6.2 ER-Based Informativeness

OVERVIEW In this section, we will first introduce weights for the edges of the

6.2. ER-Based Informativeness

underlying ER graph. These weights will be based on co-occurrence statistics for entities and relationships; they will be derived from the domain represented by the ER graph. Then, we will exploit the edge weights to compute IRank, a random-walk-based measure for capturing the informativeness of nodes in ER graphs. Finally, we will show how IRank can be extended to capture the informativeness of whole subgraphs.

BASICS Let $G = (V, l_{Ent}, E_{Rel})$ be an ER graph. In Chapter 3 (Definition 1), we introduced ER graphs as labeled multigraphs over finite sets of entity and relationship labels, which we denoted by Ent and Rel respectively. According to that definition, the labeled edges of G are $E_{Rel} \subseteq l_{Ent}(V) \times Rel \times l_{Ent}(V)$, where $l_{Ent} : V \rightarrow Ent$ is an injective function. We refer by facts to the labeled edges of G. For example, the edge (Max_Planck, citizenOf, Germany) in Figure 20 represents a fact about the entities Max_Planck and Germany.

Since the direction of a relationship between two entities can always be interpreted in the converse direction, we view the edges of an ER graph as bidirectional. That is, we assume that for each edge $(u, r, v) \in E_{Rel}$ there is an edge $(v, r^-, u) \in E_{Rel}$, where r^- represents the inverse relation label of r.

DISCUSSION We believe that in order to compute the informativeness of a node in an ER graph, the link structure has to be taken into account. On the other hand, we are aware of the fact that the edges of an ER graph do not always entail a "clear" endorsement. Consequently, measures that build on the link-based endorsement hypotheses such as PageRank [33] or HITS [103, 102] are not always applicable to ER graphs in a straight forward manner. For example, Consider an RDF database about scientists that contains for each scientist only the name, the date of birth, and the profession. Suppose that the facts (Albert_Einstein, instanceOf, Physicist) and (Bob_Unknown, instanceOf, Physicist) are contained in this database. Now, consider the respective edges in the corresponding ER graph. Since the link structure of scientist nodes in this ER graph is determined by their schema, both Albert_Einstein and Bob_Unknown will have the same link structure. Consequently, in this example, they will be endorsed equally by the link structure. Furthermore, the direction of an edge in an ER graph merely corresponds to the relationship label of that edge. Analogously, the fact (Albert_Einstein, instanceOf, Physicist) could be represented as (Physicist, hasInstance, Albert_Einstein). Hence, edge directions in an ER graph do not always reflect a "clear" endorsement.

Our informativeness measure for nodes overcomes these problems by building on edge weights that are based on co-occurrence statistics for entities and relationships. These statistics will guide a random walk process on the adjacency matrix of the ER graph. We show in the next subsection how to compute them from the domain from which the ER graph was derived.

6.2.1 Statistics-Based Edge Weights

For each fact represented by an edge, we compute two weights; one for each direction of the edge (note that we view edges as bidirectional). Each of these weights will represent a special kind of endorsement, obtained from domain-based co-occurrence statistics for entities and relationships.

DEFINITION 7: [*Fact Pattern, Match, Binding*]
Let X be a set of entity variables (placeholders for entities). A fact pattern *from an ER graph $G = (V, l_{Ent}, E_{Rel})$ is a triple $(\alpha, \beta, \gamma) \in (Ent \cup X) \times Rel \times (Ent \cup X)$, in which either $\alpha \in X$ or $\gamma \in X$, such that if $\alpha \in X$ then there is an edge (α', β, γ) in E_{Rel}, and if $\gamma \in X$ then there is an edge (α, β, γ') in E_{Rel}.*
Without loss of generality, let $\alpha \in X$. The edge (α', β, γ) from G is called a match *to the fact pattern (α, β, γ), and the entity α' is called a* binding *to the variable α.*

Consider the fact pattern (*x, instanceOf, Physicist*), $x \in X$. The fact (*Max_Planck, instanceOf, Physicist*) is a match to this pattern. In general, there may be multiple matches to a fact pattern. For example, the facts (*Albert_Einstein, instanceOf, Physicist*) and (*Bob_Unknown, instanceOf, Physicist*) could be further matches to the above fact pattern. However, as in the ranking framework of NAGA, not all matches are equally informative. In our example, the fact (*Albert_Einstein, instanceOf, Physicist*) should have a higher informativeness than (*Bob_Unknown, instanceOf, Physicist*). More precisely, the binding *Albert_Einstein* should be more informative than *Bob_Unknown*. To capture this notion of informativeness, we introduce a probabilistic model.

Let (α, β, γ) be a fact pattern, where $\alpha \in X$. Let α' be a binding of α. We estimate the informativeness of α' given the relationship β and the entity γ as:

$$P_{info}(\alpha'|\beta, \gamma) = \frac{P(\alpha', \beta, \gamma)}{P(\beta, \gamma)} \approx \frac{W(\alpha', \beta, \gamma)}{W(\beta, \gamma)} \qquad (6.1)$$

where $W(\alpha', \beta, \gamma)$ denotes the number of domain witnesses for the fact (α', β, γ), i.e., the number of occurrences of the fact (α', β, γ) in the underlying domain of the ER graph. Analogously, $W(\beta, \gamma)$ stands for the number of witnesses for the pattern $(*, \beta, \gamma)$, where the wild card '*' can be any entity. The value $P_{info}(\alpha'|\beta, \gamma)$ is assigned as a weight to the edge $\gamma \xrightarrow{\beta} \alpha'$.

We will discuss, at the end of this subsection, how $W(\alpha', \beta, \gamma)$ and $W(\beta, \gamma)$ can be estimated in practice.

To see why this formulation captures the intuitive understanding of informativeness for facts, consider the following examples. Let p =(*Albert_Einstein, instanceOf, x*) be a fact pattern, where $x \in X$. Let (*Albert_Einstein, instanceOf, Physicist*) and (*Albert_Einstein, instanceOf, Philosopher*) be two respective matches (i.e., *Physicist* and *Philosopher* are two bindings for *x*). Here, the statistics-based

6.2. ER-Based Informativeness

P_{info} value measures how often Einstein is mentioned as a physicist as compared to how often he is mentioned as a philosopher. Assuming that the underlying ER graph represents a large subset of the Web knowledge (i.e., the domain is given by the Web content), (*Albert_Einstein, instanceOf, Physicist*) is more informative than (*Albert_Einstein, instanceOf, Philosopher*), since there are more Web pages about Einstein as physicist. In this case, the statistics-based P_{info} value measures the degree to which Einstein is a physicist (or a philosopher, respectively).

Now consider the fact pattern $p = (x, instanceOf, Physicist)$ and the matches (*Albert_Einstein, instanceOf, Physicist*) and (*Bob_Unknown, instanceOf, Physicist*). That is, *Albert_Einstein* and *Bob_Unknown* are two bindings for *x*. In this case, the statistics-based P_{info} value will capture how often Einstein is mentioned as a physicist as compared to how often *Bob_Unknown* is mentioned as a physicist. Since Einstein is an important individual among the physicists, (*Albert_Einstein, instanceOf, Physicist*) will have a higher informativeness than (*Bob_Unknown, instanceOf, Physicist*). Hence, in this case, P_{info} measures the importance of Einstein in the world of physicists.

IMPLEMENTATION OF P_{info} WEIGHTS Consider the fact pattern (α, β, γ) with $\alpha \in X$. Let α' be a binding of α. In order to estimate $P_{info}(\alpha'|\beta, \gamma)$ we need to compute the numbers of witnesses $W(\alpha', \beta, \gamma)$ and $W(\beta, \gamma)$. In practice, for their estimation one can use a "background corpus", either a large Web sample, reflecting the domain of the ER graph, or the entirety of Wikipedia texts. Note that implementation-wise it is very difficult to identify all occurrences of (α', β, γ) or the occurrences of (β, γ), especially because the relationship label β can be expressed in non-trivial ways. Hence, we need to estimate the number $W(\alpha', \beta, \gamma)$ of witnesses for the fact (α', β, γ) in a more relaxed way. $W(\alpha', \beta, \gamma)$ can be estimated as the number of documents (or paragraphs, or sentences) in the background corpus in which α' and γ co-occur. $W(\beta, \gamma)$ can be estimated analogously as the number of documents containing γ. With these ingredients, we estimate $P_{info}(\alpha'|\beta, \gamma)$ as:

$$P_{info}(\alpha'|\beta, \gamma) \approx \frac{\#docs(\alpha', \gamma)}{\#docs(\gamma)} \qquad (6.2)$$

where $\#docs(\alpha', \gamma)$ stands for the number of documents in the background corpus that contain α' and γ.

Although our estimation is oblivious to relationships, it captures the intuition described above in the overwhelming majority of the cases. In our current implementation, we have precomputed the co-occurrence statistics based on inverted indexes on the Wikipedia corpus.

A seemingly simpler strategy would be to compute the co-occurrence statistics for pairs of entity names based on the hits of a Web search engine. We tried this strategy as well. It turns out that the major search engines have restrictions on the number of sequential queries posed within a certain timeframe. One can increase the number of allowed queries substantially by randomly waiting for 1 to 5 seconds before posing the next query. This way, however, it would take several months to compute the co-occurrence statistics for millions of facts.

6.2.2 IRank for Node-Based Informativeness

Our aim is an informativeness measure for nodes based on random walks on the – now weighted – ER graph. Our measure, coined *IRank* (Informativeness Rank), is related to PageRank.

PageRank [33] computes the authority of Web pages based on the link structure of the Web. In the PageRank model a random surfer walks through a directed Web graph $G(V, E)$, where V is a finite set of nodes and $E \subseteq V \times V$ is a finite set of edges. At any node $v \in V$, the surfer may continue the walk by following an outgoing edge of v with a probability inversely proportional to the out-degree of v. Alternatively, the surfer may decide to restart the walk by jumping to any random node with a probability inversely proportional to the number of nodes in G. Finally, the probability that the random surfer is at a node v is given by:

$$PR(v) = \frac{(1-q)}{|V|} + q \sum_{v' \to v} \frac{PR(v')}{O(v')} \qquad (6.3)$$

where $O(v')$ stands for the number of the outgoing edges of v', and q is a damping factor, usually set to 0.85.

The PageRank model is based on the hypothesis that every ingoing link of a Web page represents an endorsement of that Web page. However, as we have already discussed, in ER graphs the link-based endorsement hypothesis does not always hold, and consequently, methods like PageRank are not directly applicable.

Let $G = (V, l_{Ent}, E_{Rel})$ be an ER graph. Let $u \in l_{Ent}(V)$ be an entity and let $P(u)$ be the probability of encountering the entity u in the domain from which G was derived. This value can be estimated as:

$$P(u) \approx \frac{W(u)}{\sum_{v \in Ent} W(v)} \qquad (6.4)$$

where again $W(u)$ denotes the number of occurrences of the entity u in the underlying domain. $P(u)$ can be viewed as an importance prior for u.

In IRank, the random surfer may decide to restart his walk from an entity $u \in l_{Ent}(V)$ with probability proportional to $P(u)$. Alternatively, the surfer may reach u from any neighboring entity v that occurs in an edge of the form $(v, r, u) \in E_{Rel}$ (given that the surfer is at one of these neighboring entities of u).

Let $N(u)$ denote the set of neighboring entities of u in G. The probability of reaching u via one of its neighbors would be proportional to:

$$\sum_{v \in N(u)} \sum_{(v,r,u) \in E_{Rel}} P_{info}(u|r,v) \cdot IR(v) \qquad (6.5)$$

6.2. ER-Based Informativeness

where $IR(v)$ denotes the probability that the surfer is at node v, and $P_{info}(u|r,v)$ is defined as in Equation (6.1).

Finally, the accumulated informativeness at a node $u \in l_{Ent}(V)$ is given by:

$$IR(u) = (1-q)P(u) + q \sum_{v \in N(u)} \sum_{\substack{r \\ (v,r,u) \in E_{Rel}}} P_{info}(u|r,v) \cdot IR(v) \quad (6.6)$$

For practical reasons, the outgoing edge weights (i.e., the P_{info} weights) for each entity v are normalized by the sum of all outgoing edge weights of v. With this normalization step, Equation (6.6) represents an aperiodic and irreducible finite-state (i.e., an ergodic) Markov Chain. This guarantees the convergence and the stability of IRank. Although IRank is related to PageRank, the P_{info} values are crucial and make a big difference in the random walk process. In the next section, we will see that the definition of informativeness, as given by IRank (i.e., Equation (6.6)), can be modified to capture the informativeness of subgraphs that contain $k (\geq 2)$ nodes of interest from an ER graph G.

6.2.3 Most Informative Subgraphs

In this section we give an overview of our approach for estimating the informativeness of connected ER subgraphs that contain $k (\geq 2)$ entities of interest.

DEFINITION 8: [*ER Subgraph*]
Let $G = (V, l_{Ent}, E_{Rel})$ *be an ER graph. A subgraph S of G is a multigraph* $S = (V', l'_{Ent}, E'_{Rel})$*, where* $V' \subseteq V$*,* $E'_{Rel} \subseteq E_{Rel}$*, and for every edge* $(\alpha, \beta, \gamma) \in E'_{Rel}$ *there are nodes* $u, v \in V'$ *with* $l_{Ent}(u) = \alpha$ *and* $l_{Ent}(v) = \gamma$*.*

The subgraph S is connected *if for every node* $u \in V'$ *there is a node* $v \in V'$*, such that, for a relationship label* $r \in Rel$*,* $(l_{Ent}(u), r, l_{Ent}(v)) \in E'_{Rel}$ *or* $(l_{Ent}(v), r, l_{Ent}(u)) \in E'_{Rel}$

In the following, for any subgraph S of an ER graph G, we will denote by $Ent(S)$ the set of its labeled nodes (i.e., entities), and by $F(S)$ the set of its labeled edges (i.e., facts). Note that $F(S)$ contains edges of the form (α, β, γ), and that both $\alpha, \gamma \in Ent(S)$. We say a subgraph S contains an entity q if there is a labeled node $q \in Ent(S)$.

Formally, the general problem that motivates the work presented in this chapter is the following.

DEFINITION 9: [*General Problem Definition*]
Given: *an ER graph G, a set $Q = \{q_1, ..., q_k\}, k \geq 2$ of query entities, and an integer node budget $b > k$.*
Task: *find a* connected *subgraph S of G with at most b nodes that contains all entities from Q and maximizes an informativeness function $g_{info}(S, Q)$.*

6.2. ER-Based Informativeness

Intuitively, $g_{info}(S,Q)$ represents a local goodness function that increases in regions of G which contain facts that nicely capture the relations between the query entities, and decreases in regions whose facts do not contribute to the relatedness between the query entities. Given this purely intuitive nature of g_{info}, it is inherently hard to define corresponding functions. In fact, as we will see later, our approach aims to approximate an implicit g_{info} by exploiting Equation (6.6), in order to mine the most informative subgraph.

RECAPITULATION OF PREVIOUS APPROACHES A simpler version of the problem, namely for two query entities, was first introduced in [69]. The authors present an approach that emulates electrical circuits to retrieve the subgraph that best captures the relations between two given entities. The approach proceeds by determining a connected candidate subgraph C of G that contains many important interconnections between the query nodes. Then, a current-flow-based method determines the subgraph S of C that "best" connects the query nodes, i.e., the most important subgraph (with respect to the underlying current-flow-based measure).

CEPS [139] allows any number of query nodes, and addresses the problem of finding *centerpieces*, i.e., intermediate nodes that are closely connected to most of the query nodes. Random surfers exercising random walks with restarts from each query node help determining a subgraph S of G that captures the main relations between the query nodes. While [69] can efficiently deal with disk-resident graphs, CEPS is not directly applicable to them.

OUR APPROACH AT A GLANCE Following the strategy of [69], our approach, too, proceeds by generating a connected candidate subgraph C that contains all entities from Q and many important interconnections between them. The focus in this generation phase is on recall rather than on precision; that is, during this generation phase, most of the spurious regions of the graph G are removed.

The next phase aims at mining the most informative subgraph S in the generated candidate graph C that interconnects all entities from Q. Based on random walks with restarts that build on Equation (6.6), we learn for each node v in $Ent(C)$ two scores: $P_+(v)$, representing an informativeness score for v with respect to the query nodes, and $P_-(v)$, representing how uninformative v is. The label $lab(v) \in \{-,+\}$ of v is computed as $lab(v) = \arg\max_{l \in \{-,+\}} P_l(v)$ (i.e., as the label indicated by the maximum of the above two scores). That is, v is labeled $+$ if $P_+(v) \geq P_-(v)$, and $-$ if $P_+(v) < P_-(v)$. Finally, we determine a connected subgraph S of C with at most b nodes, which are all labeled '+'. Our method is designed in such a way that it guarantees the interconnection of the query nodes in the final result graph S.

Furthermore, our method has two main advantages: (1) it avoids the explicit definition of an informativeness function g_{info}, and (2) it avoids crude and non-transparent thresholding on edge and node scores in the extraction phase. Note that both methods described above (i.e., [69] and [139]) use thresholds on edge and node scores for their mining process.

6.3. The MING Algorithm

The main steps of our mining approach are the following:

1. As a first step, we apply the STAR algorithm from [95] to find a minimum-cost tree T in the generated candidate graph C that interconnects all entities from Q. In this step, the cost function for any subtree T of C that contains all query entities is given by $\sum_{e \in F(T)} d(e)$, where $d(e)$ can be any distance function that is inversely proportional to the connection strength between the two end nodes of e. Apart from being very efficient, STAR comes with a nice approximation guarantee, and experiments on real-life data sets have shown that the trees it returns are minimal in the majority of the cases. Note that the tree T determined in this step already represents a "close" relation between the entities in Q. This tree also guarantees the interconnection of all query nodes in the final graph S.

2. In a second step, each node $v \in Ent(T)$ is considered informative and is assigned the label '+'. All the nodes on the "rim" of the candidate graph C, i.e., the nodes that do not contribute to any path that interconnects query entities, are viewed as uninformative; they are assigned the label $-$. The main assumption in this step is that T already captures some relatedness between the query entities.

3. Then, for each unlabeled node $v \in Ent(C)$ and for each label $l \in \{-, +\}$ we estimate the probability $P_l(v)$ that v is visited by a random walker who starts at any node labeled l and ends up at any node labeled l. Again, we envision a random walker who is guided by the P_{info} values (see Equation (6.1)). For each node $v \in Ent(C)$ we determine its label $lab(v) = \arg\max_{l \in \{-,+\}} P_l(v)$.

4. Finally, we extract a connected subgraph S of C that contains T and has the following properties:
   - Every node in S is labeled '+',
   - S has at most b nodes,
   - S maximizes $\sum_{v \in Ent(S)} P_+(v)$.

   Note that since the initial tree T is part of the final subgraph S, it is guaranteed that all query entities are interconnected in the final result.

In the following, we discuss the details of our approach.

6.3 The MING Algorithm

Our approach, MING, consists of two main phases. Given an ER graph G and k query entities, in the first phase, MING generates a connected candidate subgraph C that contains all entities from Q and many important interconnections between them. The second phase consists in determining and extracting the most informative connected subgraph S of C that interconnects the query entities.

6.3.1 First Phase: Candidate Subgraph Generation

Our generation algorithm for the candidate subgraph C is related to the one presented in [69]. A high-level overview of our candidate generation method is given by Algorithm 6. The algorithm proceeds by applying a series of expansions starting from each node representing a query entity $q_i \in Q$. More precisely, with each query entity q_i, we associate a set of nodes $Ex(q_i)$, representing the set of already expanded nodes, and a set $Pe(q_i)$ of pending nodes, representing seen but not yet expanded nodes (lines 1, 2). In the beginning, each set $Ex(q_i)$ contains only q_i (line 1). Each set $Pe(q_i)$ contains all neighboring entity nodes of q_i in C. We denote this set by $N(q_i)$ (line 2). In each step, one of the $Ex(q_i)$ is chosen to be expanded by the node $v \in Pe(q_i)$ that is "best" connected (i.e., with respect to the P_{info} edge weights) to the nodes that are already in $Ex(q_i)$ (lines 5,8). In contrast to the extraction algorithms from [69] and [139], which use a best-first expansion strategy (i.e., in each expansion step, the most promising node is expanded), we exploit a *balanced expansion heuristics*. That is, in each step we choose the set $Ex(q_i)$ that has the lowest cardinality among the expanded sets and expand it by the node $v \in Pe(q_i)$ that is "best" connected to the nodes from $Ex(q_i)$ (lines 4,5). As shown in [82], this heuristics performs very well in practice and has satisfactory bounds on the worst case performance.

ALGORITHM 6: candidateGeneration(Q, G)
Input: ER graph G,
            set Q of query entities
Output: well-connected subgraph C that contains all entities from Q
1 Set $Ex(q_i) = \{q_i\}$ //for all $q_i \in Q$
2 Set $Pe(q_i) = N(q_i)$ //for all $q_i \in Q$
3 WHILE not stoppingCondition DO
4     $q = \arg\min_{q_j \in Q} |Ex(q_j)|$
5     $v = \arg\max_{v \in Pe(q)} \sum_{u \in Ex(q)} P_{info}(u|v) + P_{info}(v|u)$
6   expand(v)
7     $Pe(q) = Pe(q) \setminus \{v\}$
8     $Ex(q) = Ex(q) \cup \{v\}$
9     $Pe(q) = Pe(q) \cup \{u | u \in N(v), u \notin Ex(q) \cup Pe(q)\}$
10 END WHILE
11 RETURN connected subgraph C from $\cup_i (Ex(q_i) \cup Pe(q_i))$

The expansion strategy is guided by the P_{info} values, where $P_{info}(u|v)$ is defined

6.3. The MING Algorithm

as:
$$P_{info}(u|v) := \sum_{\substack{r \\ (u,r,v) \in F(G)}} P_{info}(u|r,v) \qquad (6.7)$$

and $P_{info}(u|r,v)$ is defined as in Equation (6.1). $P_{info}(v|u)$ is defined analogously.

A newly expanded node v is moved from $Pe(q)$ to $Ex(q)$, and $Pe(q)$ is updated with the neighbors of v that have not yet been seen (lines 7-9).

In analogy to the algorithm in [69], the stopping condition puts limits on the number of nodes in the intersection $\bigcap_i Ex(q_i)$ of the expanded sets. Algorithm 6 generates a candidate subgraph in $O(|Q||Ent(G)|^2)$ steps. Note that the subgraph extracted in this phase typically contains only a few thousands of nodes and edges and can be easily processed in main memory.

6.3.2 Second Phase: Mining the Most Informative ER Subgraph

Given the candidate subgraph C, we run the STAR algorithm [95] to determine a subtree T of C that closely interconnects all entities from Q. Assuming that T already captures some relatedness between the query entities, each node $v \in Ent(T)$ is viewed as informative, hence these nodes are assigned the label '+'. Nodes on the "rim" of C that do not contribute to any connection between query entities are viewed as uninformative. Consequently, they are assigned the labeled '−'.

DEFINITION 10: [*Rim Nodes*]
Let C *be a connected subgraph of G that contains all entities from Q. A rim node of C is a node that has degree one and does not represent any entity from Q.*

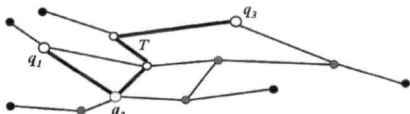

Figure 21: Sample candidate subgraph C with query nodes q_1, q_2, q_3.

Figure 21 depicts a sample candidate subgraph C. The black-colored nodes are exactly the rim nodes of C. They are labeled '−'. The bold edges in C represent the edges of the tree T returned by the STAR algorithm. The nodes of T (i.e., the white-colored nodes) are labeled '+'. The remaining nodes of C (i.e., the gray-colored nodes) remain without labels.

For each unlabeled node $v \in Ent(C)$, we compute a score $P_-(v)$, representing how uninformative v is, and a score $P_+(v)$, representing how informative v is with respect to the query entities. We will see that in our approach these two scores are not complementary. More precisely, a high $P_+(v)$ score for a node v does not necessarily imply a low $P_-(v)$ score, and vice versa. In fact, we will be merely interested in the maximum of these two scores.

In this setting, the informative subgraph mining problem can be stated as follows.

6.3. The MING Algorithm

DEFINITION 11: [*Informative Subgraph Mining*]
Given: the connected candidate subgraph C that contains all query nodes $q_1, ..., q_k \in Q, k \geq 2$, and an integer node budget $b \geq |Ent(T)|$.

Tasks:

1. *Determine for each node $v \in Ent(C)$ a label $lab(v) \in \{-,+\}$ as $lab(v) = \arg\max_{l \in \{-,+\}} P_l(v)$.*

2. *Extract a connected subgraph S of C that contains T and has the following properties: (1) every node $v \in Ent(S)$ is labeled '+', (2) S contains at most b nodes, (3) S maximizes $\sum_{v \in Ent(S)} P_+(v)$.*

Since we require that the tree T be a subgraph of S, we guarantee that all query nodes are interconnected in the result graph. In addition, we will see in Section 6.4 that T also helps constructing result graphs in which all query nodes are similarly well interconnected.

In the following, we present a classification algorithm for learning a label $l \in \{-,+\}$ for each unlabeled node of C.

CLASSIFICATION ALGORITHM The intuition behind our classification method is the following. Let $l \in \{-,+\}$. Consider all paths in C that connect any two nodes labeled l and cross an unlabeled node v. The higher the number of such paths, the higher the probability is that v is also labeled l. On the other hand, the longer these paths are, the smaller the probability is that v is labeled l. In order to estimate $P_l(v)$, we need methods that capture and reward robust structural connectivity and discourage long and loose connections.

Consider a random walker that starts at a node labeled l in C and finishes his walk again at a node labeled l. For an unlabeled node $v \in Ent(C)$, let $P_l(v)$ denote the probability that v is visited during this random walk. As depicted in Figure 22, we estimate this probability as the composition of two probabilities $P_l^1(v)$ and $P_l^2(v)$. $P_l^1(v)$ represents the probability that the random walker starts at any l-labeled node and reaches v. $P_l^2(v)$ represents the probability that any l-labeled node is reached when the random walker starts his walk at v. It is straightforward to see that $P_l(v) = P_l^1(v) \cdot P_l^2(v)$.

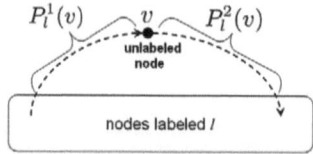

Figure 22: Probability P_l composed of the probabilities P_l^1 and P_l^2.

6.3. The MING Algorithm

In order to estimate $P_l^1(v)$, we extend IRank into a Random Walk with Restarts (RWR) process. The reason for this is the following. In a random walk process such as the one represented by PageRank or IRank (see Equation (6.6)), the steady-state probabilities of nodes are independent of the initial probability distribution on the nodes. Furthermore, long paths are not punished. In fact, long paths between nodes do not play any role in the random walk process (or the steady-state probabilities). This is different in an RWR process. There, nodes that are far away from the starting nodes will be visited less frequently, because of the restart probability. Hence, long connectivity paths are discouraged in a natural way. Furthermore, as reported in [140] and [139], RWRs have very nice properties when it comes to capturing the structural connectivity between nodes. They overcome several limitations of traditional graph distance measures such as maximum flow, shortest paths, etc.

The idea behind our RWR process is the following. The walk starts at any l-labeled node v and follows the outgoing edges of v with a probability that is proportional to the edge weights (as edge weights on C we consider the P_{info} values from Equation (6.1)). The probability that our walk follows the outgoing edges of nodes is dampened by a factor q (damping factor). With probability $(1-q)$ the random walk restarts at any node that is labeled l.

For each node $v \in Ent(C)$, let

$$pr_l(v) = \begin{cases} \frac{1}{\#\{v \in Ent(C); lab(v)=l\}}, & lab(v) = l \\ 0, & otherwise \end{cases}$$

and let $R_l := [pr_l(v)]_{v \in Ent(C)}$ be the vector representing the restart probabilities. Note that for each unlabeled node in C the restart probability is 0. Let $P := [P_l^1(v)]_{v \in Ent(C)}$ denote the *steady-state probability* vector of an RWR starting at nodes labeled l. The RWR is formally described by:

$$P = q\tilde{A}_W P + (1-q)R_l \quad (6.8)$$

where $\tilde{A}_W$ is the column-normalized, weighted adjacency matrix of the ER graph. Note that $\tilde{A}_W$ contains the normalized P_{info} values derived from the underlying domain. More precisely, the position representing the adjacent entity nodes (u, v) in $\tilde{A}_W$ is assigned the value $P_{info}(v|u)$ normalized by the sum of all outgoing edge weights of u, where $P_{info}(v|u)$ is defined analogously to Equation (6.7).

Finally, the vector P can be computed by iterating the following equation until convergence.

$$R^{i+1} = q\tilde{A}_W R^i + (1-q)R_l \quad (6.9)$$

where R^0 is set to R_l. By applying this method once for each $l \in \{-,+\}$, we can estimate for each unlabeled node v the probability $P_l^1(v)$.

In order to compute P_l^2 for an unlabeled node v, we could use the same RWR technique. More precisely, we could run an RWR for every unlabeled node v

6.3. The MING Algorithm

and compute $P_l^2(v)$ as $P_l^2(v) = \sum_{u:lab(u)=l} P_v(u)$, where $P_v(u)$ would denote the stationary probability of u as determined by the RWR starting at v. However, there might be several hundreds of unlabeled nodes in C, and running an RWR for each of the unlabeled nodes is highly inefficient in practice. Hence, we estimate the P_l^2 in a more relaxed but more efficient way.

Let u be an unlabeled node in C. The probability of having been at node u one step before reaching any node v labeled l is given by:

$$P(u,1) = \sum_{\substack{v:lab(v)=l \\ v \in N(u)}} P_{info}(v|u) \qquad (6.10)$$

where $N(u)$ denotes the set of neighboring nodes of u in C.

Let $L \subseteq Ent(C)$ denote the set of nodes labeled l in C. Now, one can recursively define the probability that u is reached $s > 1$ steps before any node labeled l as:

$$P(u,s) = \sum_{v \in Ent(C) \setminus L} P_{info}(v|u) \cdot P(v, s-1) \qquad (6.11)$$

Intuitively, s represents the depth of the recursion.

As shown in Algorithm 7, the above recursion can be computed in an iterative manner in time $O(|F(C)|)$.

ALGORITHM 7: p2lEstimation(C)
Input: ER subgraph C,
Output: estimated value of $P_l^2(v)$ for all $v \in Ent(C)$,
1   $X := \{v | lab(v) = l\}$
2   FOR EACH $v \in X$
3     $P_l^2(v) = \frac{1}{|X|}$
4   END FOR
5   $Y := \emptyset; U := Ent(C) \setminus X$
6   WHILE U is not empty DO
7   FOR EACH pair of adjacent nodes u, v with $u \in U, v \in X$
8   compute $P_l^2(u) = \sum_{v:v \in X} P_{info}(v|u) P_l^2(v)$
9   insert u into Y
10   END FOR
11    $U := U \setminus Y$
12    $X := Y; Y := \emptyset$
13 END WHILE

In lines 1 - 4 of Algorithm 7, all nodes in X (which are exactly the nodes labeled l) are assigned the same P_l^2 value $\frac{1}{|X|}$. The set U (line 5) contains in each iteration (lines 6 - 13) all unlabeled nodes that have no P_l^2 value. In each iteration, we exclude from U (line 11) all nodes for which a P_l^2 value was determined during the iteration (represented by the set Y, line 5). At the end of each iteration, the set X is set to Y. In lines 7 - 10, for each pair of adjacent nodes u, v with $u \in U$ and $v \in X$ we compute $P_l^2(u)$ (line 8). The algorithm terminates when the set U is empty.

At this point, each node v of C has for each $l \in \{-, +\}$ a probability $P_l(v) = P_l^1(v) \cdot P_l^2(v)$. The label of each node in $v \in Ent(C)$ can now be easily determined by $lab(v) = \arg\max_{l \in \{-,+\}} P_l(v)$. Finally, the most informative subgraph of C is the one that consists of all nodes v for which $lab(v) = +$. In case this subgraph has more than b nodes, we successively remove from it the node v that does not belong to T and has minimal $P_+(v)$. By the construction of our mining method, it is easy to see that S fulfills the desired properties of Definition 11.

Although the special problem addressed in this section comes with two classes of nodes (i.e., *informative* and *uninformative* nodes), our classification approach can easily be generalized to more than two classes. One would have to compute the $P_l(v)$ probabilities as described above for each class label l. This way the subgraph that best represents a certain class of nodes could be retrieved.

6.4 Experimental Evaluation

For the evaluation of MING we focused on two aspects: (1) extraction efficiency, and (2) quality of the mined subgraphs. In this section, we will present performance results of MING in comparison with the state-of-the-art approaches FSD (for Fast Subgraph Discovery) [69] and CEPS [139].

COMPETITORS In its first phase, FSD efficiently extracts a connected candidate subgraph C that contains many important connections between the query nodes. The candidate generation algorithm applies a series of expansions starting from the query nodes. The expansions follow a best-first strategy and stop when a stopping condition is fulfilled. In a second phase, a final answer graph S is mined from C. This is done by means of a current-flow-based algorithm. In contrast, the more recent approach, CEPS, extracts the most important subgraph S (that captures the main relations between the query nodes) directly from G by determining the most central nodes of G with respect to the query nodes. This is done by applying an RWR from each query node. For each node, the stationary probabilities from each RWR are multiplied to a final node score. The top-k nodes with highest scores constitute the central nodes (i.e., the *centerpieces*). To extract the final subgraph S from G, the authors propose an extraction algorithm that generalizes the candidate generation algorithm of FSD for more than two query nodes. While the first phase of MING pursues the same goal as the first phase of FSD, the second phase of MING is rather related to CEPS. All methods are implemented in Java.

EVALUATION ASPECTS As for the efficiency aspect, we have evaluated the

6.4. Experimental Evaluation

performance of MING on the task of extracting the candidate subgraph C. Therefor, we have compared the running times of our candidate subgraph generation method (represented by Algorithm 6) with the running times of a generalized FSD that works for more than two query nodes. In a second set of experiments, we have evaluated the running time of MING on the task of determining the most informative subgraph S from C. Here, we have compared the mining efficiency of MING with the mining efficiency of CEPS (for same candidate subgraphs C). All efficiency experiments were performed on a 2 GHz Pentium machine with 2 GB of main memory and an Oracle Database (version 9.1) as the underlying persistent storage.

As for the quality aspect, we have conducted an extensive user evaluation to asses the informativeness (i.e., the intuitive understanding of relatedness between given query entities) of result graphs returned by MING and CEPS.

DATA SETS As data sets we have used YAGO [137, 138, 136] and DBLP. The ER graph given by the latest version of YAGO contains more than 2 million nodes (i.e., entities) and 20 million edges (i.e., facts). YAGO combines facts extracted from Wikipedia with facts from WordNet [72]. It supports more than 100 interesting relationship labels (e.g., *hasChild, hasWonPrize, hasAcademicAdvisor, graduatedFrom, bornIn, bornOnDate, marriedTo, actedIn,* etc.), and knows the majority of the entities known to Wikipedia.

From the latest XML version of DBLP we extracted an ER graph consisting of 2 million nodes (representing authors, publications, publication types, conferences, and journals) and 9 million edges with relationship labels that describe important information about publications and authors (such as, *hasAuthor, appearedIn, publishedInYear, coAuthorOf,* and *hasPublicationType*). Apart from being sparser than the YAGO graph, the DBLP graph is also in terms of entities and relationship labels much less diverse.

Both ER graphs (i.e., the YAGO graph and the DBLP graph) are stored in a relational database with the simple schema

$$EDGE(E_1, relation, E_2, PinfoE_1E_2, PinfoE_2E_1),$$

where E_1, E_2 are entity names and $PinfoE_iE_j$ is a score approximating the value $P_{info}(E_j|relation, E_i)$ given by Equation (6.1). For YAGO these scores were estimated by means of co-occurrence statistics for entities. These statistics were directly derived from the Wikipedia corpus as described in Subsection 6.2.1.

For DBLP on the other hand, it is very difficult to find an adequate domain from which co-occurrence statistics can be derived. Hence, for the DBLP facts we assume uniform P_{info} values.

6.4.1 Efficiency

The runtime of FSD and MING is clearly dominated by the candidate subgraph extraction task. For our comparison we used two query sets, one for DBLP and

6.4. Experimental Evaluation

one for YAGO. Each set contained 30 randomly generated queries, where each query consisted of 3 entities. For both query sets, the average runtime of FSD was compared with the average runtime of our candidate extraction method. Both methods were evaluated for each query, based on the same stopping condition (see Algorithm 6). Additionally, both methods were treated uniformly as far as the overhead for database calls is concerned. The results are presented in Figure 23. The candidate generation method of MING clearly outperforms FSD's generation method. On average, MING generates a candidate subgraph in less than 10 seconds and is at least 5 times faster than FSD on both datasets. It is important to note that this runtime difference has considerable consequences for the user-perceived response time. While the perceived response time of MING is acceptable, the perceived response time of FSD is unsatisfactory.

The better runtimes of the methods on the YAGO graph can be explained through the denser structure of YAGO.

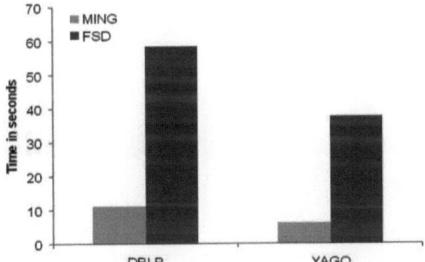

Figure 23: Avg. extraction runtimes for FSD and MING.

In a second experiment, we evaluated the performance of CEPS and MING on the task of mining the final answer graph from a given candidate subgraph. For each of the graphs (YAGO and DBLP), we randomly generated query sets of queries with 3,4,5, and 6 query nodes. Each set contained 15 queries, resulting in 60 queries per graph. For each candidate subgraph C generated by MING for each query, we measured the average time needed by MING and CEPS to mine the final subgraph S. The results are presented in Figures 24 and 25. The good runtime of CEPS for three query nodes reflects the fact that MING uses a more intricate mining technique. MING applies the STAR algorithm and two RWRs on the candidate subgraph (one from the nodes of the tree T that interconnects the query nodes in C, and one from the rim nodes of C). Although CEPS runs one RWR per query node, in the case of three query nodes the running times are comparable. However, as the number of query nodes increases, MING clearly outperforms CEPS. Note that the runtime of MING in this phase is negligible when compared to the runtime for the extraction of the candidate subgraph. Nevertheless, Figures 24 and 25 indicate the superiority of our mining method over CEPS in this phase. The worse runtimes of both methods on the DBLP graph can be explained by the fact that the subgraphs extracted from DBLP are of a higher order than the subgraphs extracted from YAGO. This leads to

6.4. Experimental Evaluation

higher runtimes for the RWR computations.

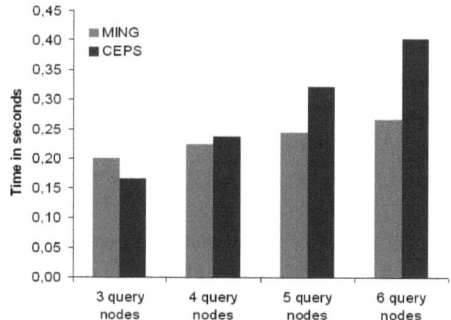

Figure 24: Avg. mining times for CEPS and MING on subgraphs from DBLP.

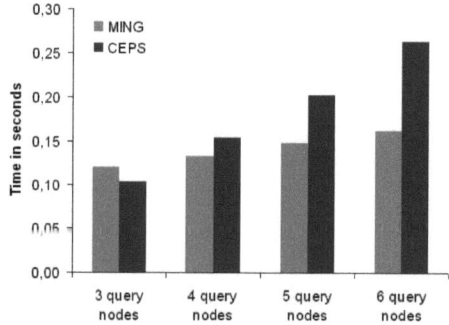

Figure 25: Avg. mining times for CEPS and MING on subgraphs from YAGO.

6.4.2 Quality

In order to evaluate the quality of returned subgraphs, we conducted a user evaluation. The result graphs of MING and CEPS were shown to human judges who had to decide which of the subgraphs better captured the intuition of relatedness for given query entities.

QUERIES In general, it is quite difficult for users to decide whether an ER graph that interconnects a given set of query entities is informative. The reason for this is threefold: (1) informativeness is an intuitive and also subjective notion, (2) a user's intuition has to be supported by the data in the underlying ER graph, and (3) a user needs to have very broad knowledge to assess the informativeness of a result graph for any set of given query nodes (especially when the query nodes represent rather obscure entities). Therefore, for this evaluation, we generated queries in which the query nodes represented famous individuals. Thanks to Wikipedia, YAGO is very rich in terms of famous individuals and contains plenty of interesting facts about them. In order to generate our queries, we extracted from the Wikipedia lists, a list

6.4. Experimental Evaluation

of famous physicists, a list of famous philosophers, and a list of famous actors. From each of these lists we randomly generated 20 queries, each of them consisting of 2 to 3 query entities, resulting in a set of 60 queries in total. The queries are presented in the appendix.

COMPARISON As ER graph for the user evaluation we chose the YAGO graph. The diversity of YAGO makes it simpler for users to assess whether a result graph captures the intuitive notion of informativeness or not. For each of the 60 queries above, we presented the results produced by CEPS and MING to human judges on a graph-visualization Web interface, without telling them which method produced which graph. Note that none of the judges was familiar with the project. In the visualization interface, we used the same visualization features for both methods. For visualization purposes, the result graphs of CEPS and MING were pruned, whenever they had more than 15 nodes. By restricting the result graphs to such a small number of nodes, both methods were challenged to maintain only the most important nodes in the result graphs. CEPS comes with its own pruning parameter (i.e., visualization parameter). For each query, the users were given the possibility to decide which of the presented subgraphs they perceived as more informative. That is, one of the results could be marked *informative*. We also allowed users to mark both result graphs as *informative*, if they perceived them both as equally informative. Additionally, the results of both methods could be left unmarked, meaning that they both did not suit the user's intuition. The results are presented in Table 9.

	MING	CEPS
# times preferred over competitor	182	4
# times marked *informative*	185	7
# times both marked *informative*	3	
# times both left unmarked	21	

Table 9: Results of the user evaluation

RESULTS There were 210 assessments in total, corresponding to more than 3 assessments per query. The result graphs produced by MING were marked 185 times as informative, and out of these, 182 times, they were perceived more informative than the results produced by CEPS. On the other hand, the MING results were left 25 times unmarked, and out of these, only 4 times they were perceived to be less informative than the results produced by CEPS. The results of both methods were perceived in 3 cases as equally informative, and in 21 cases equally uninformative.

The fundamental factor for the qualitative superiority of MING is its subgraph learning method. It learns informative and structurally robust paths between the nodes of an initial tree T that closely interconnects the query nodes. For this, it exploits random walks with restarts guided by co-occurrence statistics derived from the underlying domain. To illustrate the main difference between CEPS and MING, we depict in Figure 26 the answers produced by MING and CEPS for the query that

6.4. Experimental Evaluation

asks for the relations between the *Jessica Lange*, *Robert Redford*, and *Sally Field*. In this example, the result graphs were both restricted to 8 nodes. Note that restricting the result graphs to such a small number of nodes, forces both methods to maintain only the most important nodes in their results (i.e., the nodes with the highest scores). The result graph of MING (the uppermost graph in Figure 26) has identified the path that connects Sally Field and Jessica Lange through the Academy Award as informative. Furthermore, it has also identified the path that connects Sally Field and Robert Redford through the node labeled "Californian_actor" as informative. These are both findings that are missed by CEPS.

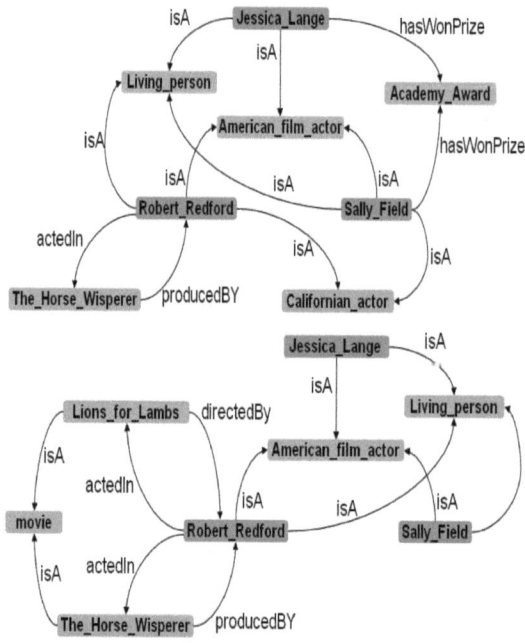

Figure 26: Answer graphs produced by MING (above) and CEPS (below).

As observed in our experiments, one of the shortcomings of CEPS is that the quality of its result graphs degrades if some of the query nodes occur in dense regions of the underlying ER graph. In this case, the result graphs become skewed towards the denser regions, especially when the number of result nodes is restricted to a small number. The node representing Robert Redford occurs in a dense region of the YAGO graph, reflecting the fact that Robert Redford has acted in several movies that were produced or directed by him. Consequently, a considerable amount of the RWR starting from this node is absorbed by this region. This leads to a skewed result graph that overemphasizes facts on individual query entities and misses salient relations between the entities. MING, on the other hand, avoids skewed result graphs by running an RWR from the nodes of the tree returned by the STAR algorithm. In our example, the node labeled "American_film_actor" is part of this tree, and contributes equally to the informativeness of nodes in the neighborhood as the query nodes. This way, MING manages to capture the informative relations that Robert Redford and Sally Field are from California, that Jessica Lange and Sally Field are both Academy Award winners, and that all three actors are alive.

These results fortify our assumption that MING indeed captures the intuitive notion of informativeness, as described in this paper, in most of the cases.

6.5 Conclusion

The motivation for this work has been to provide new techniques for exploring and discovering knowledge in large entity-relationship graphs. The presented method, MING, is a significant step forward in this realm. It contributes to new semantic measures for the relatedness between entities. MING exploits such measures for extracting informative subgraphs that connect two or more given entities. Our experimental studies have shown that MING is not only more efficient than prior approaches to this problem, but also produces outputs that are considered more informative by end-users.

A promising research direction is the integration of user interests and background knowledge into the knowledge discovery process. Social network applications dealing with user-oriented recommendation could widely profit from this challenging but exciting direction.

7 Conclusion

This work has presented techniques for querying, exploring and discovering knowledge in large knowledge bases that organize information as ER graphs. With NAGA we have provided a new framework for systems aiming at expressive search and ranking capabilities with entities and relationships. The two presented techniques STAR and MING contribute to more advanced forms of knowledge discovery on graph-structured data.

We are witnessing a strong momentum in knowledge-sharing communities, knowledge-base development, social networks and interoperability across different networks, integration of different kinds of biological networks, and other exciting trends towards a richer knowledge society. Thus, we believe that our work fills an important need.

There are various ways to extend the work presented in this work. The NAGA system could be extended into a full-fledged question answering system. For this, a natural-language-processing and pattern-matching component would have to be added on top of NAGA's query answering component. The translation of natural language questions into formal, graph-based queries is certainly challenging, but with NAGA's rich query model we already have a cornerstone for accomplishing this goal.

NAGA's framework and its techniques could be extended to better capture the context of the user and the data. User context requires personalized and task-specific search, ranking, and knowledge discovery techniques. These techniques should consider the user's interests and background knowledge, as well as the current location, time, short-term history, and intentions in the user's digital traces. Data context calls for search and ranking models that can deal with complex entity-relationship patterns beyond simple facts (edges between entities)[1].

Evaluating complex query predicates over large ER graphs is computationally hard, especially when ranking is needed. One should aim at efficient top-k techniques that avoid materializing overly large numbers of results.

The envisioned path towards Web-scale knowledge bases with efficient and expressive search, ranking, and knowledge discovery capabilities may take a long time to mature. In any case, it is an exciting challenge that should appeal to and benefit from several research communities such as Databases, Information Retrieval, Information Extraction, Natural Language Processing, Social and Semantic Web, Artificial Intelligence, and many more.

---

[1] More complex patters are for example facts holding between facts (e.g., fact A is older than fact B)

8 Appendix

8.1 Queries for the User Evaluation of NAGA

For the user evaluation of NAGA, we determined 55 questions from the question answering datasets of TREC 2005 and TREC 2006 that could be expressed by NAGA relations. The questions are shown below.

1.	When was George Foreman born?	(George_Foreman, bornOnDate, $z)
2.	When was Kurosawa born?	(Kurosawa, bornOnDate, $z)
3.	What was Kurosawa's profession?	(Kurosawa, type subClassOf, $z)
4.	What was the profession Kurosawa's wife?	(Kurosawa, isMarriedTo, $y) ($y, type subClassOf, $z)
5.	What were some of Kurosawa's Japanese film titles?	(Kurosawa, directed|produced|created, $z)
6.	What was Kurosawa's English nickname?	($z, means, Kurosawa)
7.	Name some movies that starred Paul Newman?	(Paul_Newman, actedIn, $x)
8.	Provide a list of names or identifications given to meteorites?	($y, isa, meteorite)
9.	When was the American Legion founded?	(American_Legion, establishedOnDate, $z)
10.	When was Enrico Fermi born?	(Enrico_Fermi, bornOnDate, $z)
11.	When did Enrico Fermi die?	(Enrico_Fermi, diedOnDate, $z)
12.	What was the vocation of Rachel Carson?	(Rachel_Carson, type subClassOf, $z)
13.	What books did Rachel Carson write?	(Rachel_Carson, wrote|created, $z)
14.	When did Rachel Carson die?	(Rachel_Carson, diedOnDate, $z)
15.	Of what country is Vicente Fox president?	(Vicente_Fox, politicianOf, $z)
16.	When was Vicente Fox born?	(Vicente_Fox, bornOnDate, $z)
17.	What is OPEC?	(OPEC, type subClassOf, $z)
18.	What is NATO?	(NATO, type subClassOf, $z)
19.	When was Rocky Marciano born?	(Rocky_Marciano, bornOnDate, $z)
20.	List the record titles by Counting Crows.	(Counting_Crows, created, $z)
21.	When was Woody Guthrie born?	(Woody_Guthrie, bornOnDate, $z)
22.	What year did Woody Guthrie die?	(Woody_Guthrie, diedOnDate, $z)
23.	What was the profession of Bing Crosby?	(Bing_Crosby, type subClassOf, $z)
24.	What movies did Bing Crosby act in?	(Bing_Crosby, actedIn, $z)
25.	What were some of Paul Revere's occupations?	(Paul_Revere, type subClassOf, $z)
26.	When was Paul Revere born?	(Paul_Revere, bornOnDate, $z)
27.	When did Paul Revere die?	(Paul_Revere, diedOnDate, $z)
28.	List various occupations of Jesse Ventura.	(Jesse_Ventura, type subClassOf, $z)

Table 10: Questions from TREC 2005

8.1. Queries for the User Evaluation of NAGA

1.	What is LPGA?	(LPGA, type subClassOf, $z)
2.	In what year was Warren Moon born?	(Warren_Moon, bornOnDate, $z)
3.	In what country is Luxor?	(Luxor, locatedIn*, $z) ($z, type, country)
4.	When was NASCAR founded?	(NASCAR, establishedOnDate, $z)
5.	When was Mozart born?	(Mozart, bornOnDate, $z)
6.	What is IMF?	(IMF, type subClassOf, $z)
7.	What movies did Judi Dench play in?	(Judi_Dench, actedIn, $z)
8.	In what county was Stonehenge built?	(Stonehenge, locatedIn*, $z)
9.	Which movies did Hedy Lamarr appear in?	(Hedy_Lamarr, actedIn, $z)
10.	What did Hedy Lamarr invent?	(Hedy_Lamarr, discovered, $z)
11.	What is ETA?	(ETA, type subClassOf, $x)
12.	In what state is Johnstown?	(Johnstown, locatedIn, $z)
13.	Where was Shakespeare born?	(Shakespeare, bornIn, $z) ($z, locatedIn*, $y)
14.	When was Shakespeare born?	(Shakespeare, bornOnDate, $z)
15.	When was Hitchcock born?	(Hitchcock, bornOnDate, $z)
16.	What movies did Meg Ryan star in?	(Meg_Ryan, actedIn, $z)
17.	Who was Meg Ryan married to?	(Meg_Ryan, marriedTo, $z)
18.	What government position did Janet Reno have	(Janet_Reno, type subClassOf, $z)
19.	In which movies did Frank Sinatra appear?	(Frank_Sinatra, actedIn, $z)
20.	What year was Wal-Mart founded?	(Wal-Mart, establishedOnDate, $z)
21.	What are the titles of songs written by John Prine?	(John_Prine, created|wrote, $z)
22.	Who was Carolyn Bessette-Kennedy married to?	(Carolyn_Bessette-Kennedy, isMarriedTo, $z)
23.	What songs did Patsy Cline record?	(Patsy_Cline, created, $z)
24.	Where was Cole Porter born?	(Cole_Porter, bornIn, $z) ($z, locatedIn*, $y)
25.	Name supporting actors who performed in Cheers.	($z, actedIn, Cheers)
26.	What year was Heinz Ketchup introduced?	("Heinz Ketchup", establishedOnDate, $x)
27.	What abbreviation is the International Rowing Federation also known by?	($x, means, International_Rowing_Federation)

Table 11: Questions from TREC 2006

12 questions were obtained from the work on SphereSearch [77], where a set of 50 natural language questions is provided. Again, we determined those questions that can be expressed with NAGA relations.

1.	What is the given name of the politician Rice?	(Rice, familyNameOf, $y) ($y, isa, politician) ($z, givenNameOf, $y)
2.	List movies directed by Madonna's husband.	($x, isMarriedTo, Madonna) ($x, directed, $y)
3.	List French mathematicians of the 18th century.	($x, type, french_mathematician) ($x, bornOnDate, $y) ($y, before, '1800-00-00') ($y, after, '1700-00-00')
4.	Which composers have been composing in the first half of the 18th Century?	($y, isa, composer) ($y, bornOnDate, $x) ($x, after, '1700-00-00') ($x, before, '1750-00-00')
5.	List Russian composers.	($x, type, russian_composer)
6.	Which governor acted in a science fiction movie?	($x, type, science_fiction_film) ($y, actedIn, $x) ($y, isa, governor)
7.	In which movies did a governor act?	($y, isa, governor) ($y, actedIn, $z)
8.	Which Australian singer acted in "Moulin Rouge"?	($x, actedIn, "Moulin Rouge") ($x, isa, singer) ($x, (isCitizenOf|livesIn|bornIn)locatedIn*, Australia)

Table 12: SSearch Questions

8.2. MING Queries for the User Evaluation

9.	List German physicists of the 20th century who immigrated to U.S.	($x, type, german_physicist) ($x, livesIn|isCitizenOf, United_States) ($x, bornOnDate, $y) ($y, after, '1870-00-00') ($y, before, '1970-00-00')
10.	List physicists of the 20th century who won the Nobel Prize.	($x, type, physicist) ($x, bornOnDate, $y) ($y, after, '1870-00-00') ($y, before, '1970-00-00') ($x, hasWonPrize, Nobel_Prize_in_Physics)
11.	List organizations were involved in the Watergate scandal.	($x, type, organization) ($x, context, Watergate_scandal)
12.	Which movies starred a James-Bond actor?	($x, type, James_Bond_film) ($y, actedIn, $x) ($x, actedIn, $z)

Table 13: SSearch Questions

We also constructed 18 natural language questions that can be translated into regular-expression queries.

1.	Which person by the name of Curie has won a prize?	("Curie", familyNameOf hasWonPrize, $x)
2.	Who was Pulitzer and what was his profession?	("Pulitzer", familyNameOf type subClassOf*, $x)
3.	List actors, directors or producers of James-Bond films.	($x, type, James_Bond_film) ($x, actedIn|produced|directed, $y)
4.	List movies starring an actress called Julia?	(Julia, givenNameOf actedIn, $x)
5.	Who produced or directed "Around the world in 80 days"?	($x, produced|directed, "Around the world in 80 days")
6.	List movies directed by or starring an actor named Douglas.	("Douglas", (givenNameOf|familyNameOf) (actedIn|directed), $x)
7.	List movies in which Willis was involved.	("Willis", familyNameOf (actedIn|directed|produced), $x)
8.	Where is the Rebmann Glacier located?	(Rebmann_Glacier, locatedIn*, $x)
9.	List some lakes located in Africa.	($x, isa, lake) ($x, locatedIn*, Africa)
10.	What connects Max Planck and Richard Feynman?	(Max_Planck, connect, Richard_Feynman)
11.	What do Niels Bohr and Albert Einstein have in common?	(Niels_Bohr, connect, Albert_Einstein)
12.	What connects John Gotti and Al Capone?	(John_Gotti, connect, Al_Capone)
13.	What connects Indira Gandhi and Margaret Thatcher?	(Indira_Gandhi, connect, Margaret_Thatcher)
14	What connects the musicians Michael Jackson and Prince?	(Michael_Jackson, connect, Prince_(musician))
15.	What connects the Hudson River and Black River?	("Hudson River", connect, "Black River")
16.	What do Albania and Greece have in common?	(Albania, connect, Greece)
17.	What connects Paris and Athens?	(Paris, connect, Athens)
18.	What connect Saint Helena and the Cayman Islands?	(Saint_Helena, connect, Cayman_Islands)

Table 14: OWN Questions

8.2 MING Queries for the User Evaluation

In order to generate queries for the user evaluation, we extracted from the Wikipedia lists, a list of famous physicists, a list of famous philosophers, and a list of famous actors. From each of these lists we randomly generated 20 queries, each of them consisting of 2 to 3 query entities, resulting in a set of 60 queries in total. The queries are presented in the following.

1. Paul_Dirac — Enrico_Fermi — Max_Born

8.2. MING Queries for the User Evaluation

2. Max_Planck — James_Clerk_Maxwell — Niels_Bohr

3. Richard_Feynman — Michael_Faraday — Ernest_Rutherford

4. Louis_de_Broglie — Max_Born — Michael_Faraday

5. Niels_Bohr — Ernest_Rutherford — Max_Born

6. Isaac_Newton — James_Clerk_Maxwell — Werner_Heisenberg

7. James_Clerk_Maxwell — Niels_Bohr — Stephen_Hawking

8. Werner_Heisenberg — Enrico_Fermi — Paul_Dirac

9. Max_Planck — Werner_Heisenberg — Enrico_Fermi

10. Niels_Bohr — Michael_Faraday — Max_Born

11. Edwin_Hubble — Albert_Einstein

12. Stephen_Hawking — Johannes_Kepler

13. Werner_Heisenberg — Nicolaus_Copernicus

14. Ernest_Rutherford — Blaise_Pascal

15. Hideki_Yukawa — Max_Planck

16. James_Clerk_Maxwell — Hideki_Yukawa

17. Albert_Einstein — Wolfgang_Pauli

18. Ernest_Rutherford — Johannes_Kepler

19. Ludwig_Boltzmann — Richard_Feynman

20. Isaac_Newton — Edmond_Halley

21. Val_Kilmer — Kristin_Davis — Josh_Hartnett

22. Pam_Grier — Matt_Damon — Sharon_Stone

23. Tom_Sizemore — Al_Pacino — Jennifer_Garner

24. Harrison_Ford — Robert_Redford — Sally_Field

25. Sandra_Bullock — Jennifer_Aniston — Kevin_Spacey

26. Michael_Douglas — Billy_Bob_Thornton — Kim_Delaney

27. Sigourney_Weaver — Winona_Ryder — Michael_Keaton

8.2. MING Queries for the User Evaluation

28. Sarah_Michelle_Gellar — Salma_Hayek — Viggo_Mortensen
29. Gina_Gershon — Michael_Douglas — Brittany_Murphy
30. Jessica_Lange — Sally_Field — Robert_Redford
31. Jeanne_Tripplehorn — Jennifer_Aniston — Diane_Lane
32. Clint_Eastwood — Helen_Hunt — Edie_Falco
33. Liv_Tyler — Dennis_Quaid — Teri_Hatcher
34. Demi_Moore — Ashton_Kutcher — Bruce_Willis
35. Jessica_Alba — Leonardo_DiCaprio — Billy_Crystal
36. Maria_Bello — Michael_Douglas — Uma_Thurman
37. George_Clooney — Liam_Neeson — Jake_Gyllenhaal
38. Uma_Thurman — Jake_Gyllenhaal — Jennifer_Garner
39. Kevin_Spacey — Halle_Berry — Julia_Roberts
40. Jodie_Foster — Teri_Hatcher — Christina_Ricci
41. Max_Weber — Georg_Wilhelm_Friedrich_Hegel — Ernst_Mach
42. Rudolf_Carnap — Thomas_Abbt — Max_Horkheimer
43. Johann_Gottfried_Herder — Plato — Gottfried_Leibniz
44. Arthur_Schopenhauer — Moritz_Schlick — Ludwig_Wittgenstein
45. Plato — Friedrich_Nietzsche — Bertrand_Russell
46. Ernst_Mach — Edmund_Husserl — Adam_Smith
47. Plato — Blaise_Pascal — Gottlob_Frege
48. Max_Horkheimer — Arthur_Schopenhauer — Heinrich_Hertz
49. Adam_Smith — Johann_Gottlieb_Fichte — Karl_Wilhelm_Friedrich_Schlegel
50. Max_Horkheimer — Blaise_Pascal — Bernard_Bolzano
51. Karl_Marx — Jean-Paul_Sartre — Ludwig_Wittgenstein
52. Bertrand_Russell — Albert_Einstein
53. Georg_Wilhelm_Friedrich_Hegel — Heinrich_Hertz

8.2. MING Queries for the User Evaluation

54. Arthur_Schopenhauer — Karl_Marx
55. Adam_Smith — Georg_Wilhelm_Friedrich_Hegel
56. Albert_Einstein — Edmund_Husserl
57. Johann_Augustus_Eberhard — Friedrich_Nietzsche
58. Gottlob_Frege — Bernard_Bolzano
59. Karl_Wilhelm_Friedrich_Schlegel — Karl_Marx
60. Albert_Einstein — Friedrich_Nietzsche

Bibliography

[1] Answers.com. http://www.answers.com/. Accessed 01-June-2009.

[2] DBLife. http://dblife.cs.wisc.edu/. Accessed 01-June-2009.

[3] flickr. http://www.flickr.com/. Accessed 01-June-2009.

[4] Freebase: a social database about things you know and love. http://www.freebase.com/. Accessed 01-June-2009.

[5] The gene ontology. http://www.geneontology.org/. Accessed 01-June-2009.

[6] Hakia: semantic search. http://www.hakia.com/. Accessed 01-June-2009.

[7] Jena a semantic web framework for java. http://jena.sourceforge.net/. Accessed 01-June-2009.

[8] NAGA: searching and ranking knowledge. http://www.mpi-inf.mpg.de/yago-naga/naga/demo.html. Accessed 01-June-2009.

[9] Powerset. http://www.powerset.com/. Accessed 01-June-2009.

[10] START: natural language question answering system. http://start.csail.mit.edu/. Accessed 01-June-2009.

[11] TextRunner search. http://www.cs.washington.edu/research/textrunner/. Accessed 01-June-2009.

[12] True Knowledge: the internet answer engine. http://www.trueknowledge.com/. Accessed 01-June-2009.

[13] True Knowledge: the internet answer engine, technology. http://www.trueknowledge.com/technology. Accessed 01-June-2009.

[14] Unified medical language system. http://www.nlm.nih.gov/research/umls/. Accessed 01-June-2009.

[15] Wolfram alpha: computational and knowledge engine. http://www.wolframalpha.com/. Accessed 01-June-2009.

[16] WordNet: a lexical database for the english language. http://wordnet.princeton.edu/. Accessed 01-June-2009.

[17] The YAGO-NAGA project: harvesting, searching, and ranking knowledge from the web. http://www.mpi-inf.mpg.de/yago-naga/. Accessed 01-June-2009.

[18] E. Agichtein. Scaling information extraction to large document collections. *IEEE Data Engineering Bulletin*, 28(4):3–10, 2005.

[19] E. Agichtein and S. Sarawagi. Scalable information extraction and integration, Tutorial. In *the 12th ACM SIGKDD International Conference on Knowledge Discovery and Data Mining (KDD)*, New York, NY, USA, 2006. ACM.

[20] S. Agrawal, S. Chaudhuri, and G. Das. DBXplorer: A system for keyword-based search over relational databases. In *the Proceedings of the 18th International Conference on Data Engineering (ICDE)*, pages 5–16, Washington, DC, USA, 2002. IEEE Computer Society.

[21] S. Amer-Yahia and J. Shanmugasundaram. XML full-text search: challenges and opportunities, Tutorial. In *the 31st International Conference on Very Large Data Bases (VLDB)*. VLDD Endowment, 2005.

[22] R. Angles and C. Gutierrez. The expressive power of SPARQL. In *the Proceedings of the International Semantic Web Conference (ISWC)*, Lecture Notes in Computer Science, pages 114–129, Berlin / Heidelberg, 2008. Springer.

[23] K. Anyanwu, A. Maduko, and A. Sheth. SPARQ2L: towards support for subgraph extraction queries in RDF databases. In *the Proceedings of the 16th international conference on World Wide Web (WWW)*, pages 797–806, New York, NY, USA, 2007. ACM.

[24] S. Auer, C. Bizer, G. Kobilarov, J. Lehmann, and Z. Ives. DBpedia: A nucleus for a web of open data. In *The Semantic Web*, Lecture Notes in Computer Science, pages 722–735, Berlin / Heidelberg, 2007. Springer.

[25] M. Banko and O. Etzioni. Strategies for lifelong knowledge extraction from the web. In *the Proceedings of the 4th international conference on Knowledge capture (K-CAP)*, pages 95–102, New York, NY, USA, 2007. ACM.

[26] C. D. Bateman, C. S. Helvig, G. Robins, and A. Zelikovsky. Provably good routing tree construction with multi-port terminals. In *the Proceedings of the 1997 international symposium on Physical design (ISPD)*, pages 96–102, New York, NY, USA, 1997. ACM.

[27] E. Behrends, O. Fritzen, and W. May. Querying along XLinks in XPath/XQuery: situation, applications, perspectives. In *the Proceedings of Query Languages and Query Processing, Munich, Germany (30th–31st March 2006)*, Lecture Notes in Computer Science, Berlin / Heidelberg, 2006. Springer.

[28] G. Bhalotia, A. Hulgeri, C. Nakhe, S. Chakrabarti, and S. Sudarshan. Keyword searching and browsing in databases using banks. pages 431–440, Los Alamitos, CA, USA, 2002. IEEE Computer Society.

[29] R. Bin Muhammad. A parallel local search algorithm for euclidean steiner tree problem. In *the Proceedings of the Seventh ACIS International Conference on Software Engineering, Artificial Intelligence, Networking, and Parallel/Distributed Computing (SNPD-SAWN)*, pages 157–164, Washington, DC, USA, 2006. IEEE Computer Society.

[30] C. Bizer, T. Heath, K. Idehen, and T. Berners-Lee. Linked data on the web (ldow 2008). In *Workshop at the 17th International World Wide Web Conference*, New York, NY, USA, 2008. ACM.

[31] C. Botev, S. Amer-Yahia, and J. Shanmugasundaram. A TeXQuery-based XML full-text search engine. In *the Proceedings of the 24th ACM SIGMOD international conference on Management of data*, pages 943–944, New York, NY, USA, 2004. ACM.

[32] U. Brandes, M. Gaertler, and D. Wagner. Experiments on graph clustering algorithms. In *Algorithms–ESA 2003*, Lecture Notes in Computer Science, pages 568–579, Berlin / Heidelberg, 2003. Springer.

[33] S. Brin and L. Page. The anatomy of a large-scale hypertextual web search engine. *Computer Networks and ISDN Systems*, 30(1-7):107–117, 1998.

[34] J. Broekstra, A. Kampman, and F. van Harmelen. Sesame: A generic architecture for storing and querying RDF and RDF schema. In *the Proceedings of the 1st International Semantic Web Conference on The Semantic Web (ISWC)*, pages 54–68, London, UK, 2002. Springer.

[35] M. J. Cafarella, C. Re, D. Suciu, and O. Etzioni. Structured querying of web text data: a technical challenge. In *the Proceedings of the 3rd Biennial Conference on Innovative Data Systems Research (CIDR)*, pages 225–234. www.crdrdb.org.

[36] P. Castro, S. Melnik, and A. Adya. ADO.NET entity framework: raising the level of abstraction in data programming. In *the Proceedings of the 27th ACM SIGMOD international conference on Management of data*, pages 1070–1072, New York, NY, USA, 2007. ACM.

[37] S. Ceri, G. Gottlob, and L. Tanca. What you always wanted to know about datalog (and never dared to ask). *IEEE Trans. on Knowl. and Data Eng.*, 1(1):146–166, 1989.

[38] S. Chakrabarti. Dynamic personalized pagerank in entity-relation graphs. In *the Proceedings of the 16th international conference on World Wide Web (WWW)*, pages 571–580, New York, NY, USA, 2007. ACM.

[39] M. Charikar and C. Chekuri. Approximation algorithms for directed steiner problems. *J. Algorithms*, 33(1):73–91, 1999.

[40] P. P.-S. Chen. The entity-relationship model–toward a unified view of data. *ACM Transactions on Database Systems*, 1(1):9–36, 1976.

[41] T. Cheng and K. C.-C. Chang. Entity search engine: Towards agile best-effort information integration over the web. In *the Proceedings of the 3rd Biennial Conference on Innovative Data Systems Research (CIDR)*, pages 108–113. www.crdrdb.org, 2007.

[42] T. Cheng, X. Yan, and K. C.-C. Chang. Entityrank: searching entities directly and holistically. In *the Proceedings of the 33rd international conference on Very large data bases (VLDB)*, pages 387–398. VLDB Endowment, 2007.

[43] E. F. Codd. A relational model of data for large shared data banks. *Communications of the ACM (CACM)*, 26(1), 1983.

[44] S. Cohen, Y. Kanza, B. Kimelfeld, and Y. Sagiv. Interconnection semantics for keyword search in XML. In *the Proceedings of the 14th ACM international conference on Information and knowledge management (CIKM)*, pages 389–396, New York, NY, USA, 2005. ACM.

[45] S. Cohen, J. Mamou, Y. Kanza, and Y. Sagiv. XSEarch: a semantic search engine for XML. In *the Proceedings of the 29th international conference on Very large data bases (VLDB)*, pages 45–56. VLDB Endowment, 2003.

[46] W. W. W. Consortium. The extensible markup language (XML). http://www.w3.org/TR/REC-xml/. Accessed 01-June-2009.

[47] W. W. W. Consortium. W3c: World wide web consortium. http://www.w3.org/. Accessed 01-June-2009.

[48] W. W. W. Consortium. The W3C XML path language (XPath). http://www.w3.org/TR/xpath20/. Accessed 01-June-2009.

[49] W. W. W. Consortium. The W3C XML query (XQuery). http://www.w3.org/TR/xquery/. Accessed 01-June-2009.

[50] W. W. W. Consortium. The XML linking language (XLink). http://www.w3.org/TR/xlink/. Accessed 01-June-2009.

[51] W. W. W. Consortium. The XML pointer language (XPointer). http://www.w3.org/TR/WD-xptr. Accessed 01-June-2009.

[52] W. W. W. Consortium. XQuery and XPath full text. http://www.w3.org/TR/2009/CR-xpath-full-text-10-20090709/. Accessed 01-June-2009.

[53] W. W. W. Consortium. OWL. http://www.w3.org/2004/OWL/, 2004. Accessed 01-June-2009.

[54] W. W. W. Consortium. The SPARQL language. http://www.w3.org/TR/rdf-sparql-query/, 2005. Accessed 01-June-2009.

[55] H. Cunningham. Information extraction, automatic. *Encyclopedia of Language and Linguistics, 2nd Edition*, 5:665–677, November 2006.

[56] Cycorp. Overview of OpenCyc. http://www.cyc.com/cyc/opencyc. Accessed 01-June-2009.

[57] B. B. Dalvi, M. Kshirsagar, and S. Sudarshan. Keyword search on external memory data graphs. *the Proceedings of the VLDB Endowment*, 1(1):1189–1204, 2008.

[58] N. Dalvi and D. Suciu. Efficient query evaluation on probabilistic databases. *The VLDB Journal*, 16(4):523–544, 2007.

[59] P. DeRose, W. Shen, F. Chen, A. Doan, and R. Ramakrishnan. Building structured web community portals: a top-down, compositional, and incremental approach. In *the Proceedings of the 33rd international conference on Very large data bases (VLDB)*, pages 399–410. VLDB Endowment, 2007.

[60] I. S. Dhillon, S. Mallela, and D. S. Modha. Information-theoretic co-clustering. In *the Proceedings of the 9th ACM SIGKDD international conference on Knowledge discovery and data mining (KDD)*, pages 89–98, New York, NY, USA, 2003. ACM.

[61] B. Ding, J. X. Yu, S. Wang, L. Qin, X. Zhang, and X. Lin. Finding top-k min-cost connected trees in databases. In *the Proceedings of the 23rd International Conference on Data Engineering (ICDE)*, pages 836–845, Los Alamitos, USA, 2007. IEEE Computer Society.

[62] J.-P. Dittrich and M. A. V. Salles. iDM: a unified and versatile data model for personal dataspace management. In *the Proceedings of the 32nd*

international conference on Very large data bases (VLDB), pages 367–378. VLDB Endowment, 2006.

[63] X. Dong and A. Y. Halevy. A platform for personal information management and integration. In *the Proceedings of 2nd of the International Conference on Innovative Data Systems Research (CIDR)*, pages 119–130. www.cidrdb.org, 2005.

[64] S. M. v. Dongen. *Graph Clustering by Flow Simulation*. PhD thesis, University of Utrecht, 2000.

[65] S. Dreyfus and R. Wagner. The steiner problem in graphs. *Journal of Networks*, 1:195–207, 1972.

[66] S. Elbassuoni, M. Ramanath, M. Sydow, and G. Weikum. Language-model-based ranking for queries on RDF-graphs. In *the Proceedings of the 18th ACM Conference on Information and Knowledge Management (CIKM)*, New York, NY, USA, 2009. ACM.

[67] O. Etzioni, M. Banko, S. Soderland, and D. S. Weld. Open information extraction from the web. *Communications of the ACM (CACM)*, 51(12):68–74, 2008.

[68] O. Etzioni, M. Cafarella, D. Downey, S. Kok, A.-M. Popescu, T. Shaked, S. Soderland, D. S. Weld, and A. Yates. Web-scale information extraction in knowitall: (preliminary results). In *the Proceedings of the 13th international conference on World Wide Web (WWW)*, pages 100–110, New York, NY, USA, 2004. ACM.

[69] C. Faloutsos, K. S. McCurley, and A. Tomkins. Fast discovery of connection subgraphs. In *the Proceedings of the tenth ACM SIGKDD international conference on Knowledge discovery and data mining (KDD)*, pages 118–127, New York, NY, USA, 2004. ACM.

[70] H. Fang and C. Zhai. Probabilistic models for expert finding. In *Advances in Information Retrieval, 29th European Conferenceon IR Research (ECIR)*, Lecture Notes in Computer Science, pages 418–430, Berlin / Heidelberg, 2007. Springer.

[71] O. Faroe, D. Pisinger, and M. Zachariasen. Local search for final placement in VLSI design. In *the Proceedings of the 2001 IEEE/ACM international conference on Computer-aided design (ICCAD)*, pages 565–572, Piscataway, NJ, USA, 2001. IEEE Press.

[72] C. Fellbaum. *WordNet: an Electronic Lexical Database*. MIT Press, 1998.

Bibliography

[73] D. Gibson, J. Kleinberg, and P. Raghavan. Inferring web communities from link topology. In *the Proceedings of the 9th ACM conference on Hypertext and hypermedia : links, objects, time and space—structure in hypermedia systems (HYPERTEXT)*, pages 225–234, New York, NY, USA, 1998. ACM.

[74] D. Gibson, R. Kumar, and A. Tomkins. Discovering large dense subgraphs in massive graphs. In *the Proceedings of the 31st international conference on Very large data bases (VLDB)*, pages 721–732. VLDB Endowment, 2005.

[75] M. Girvan and M. E. Newman. Community structure in social and biological networks. *the Proceedings of the National Academy of Sciences of the United States of America*, 99(12):7821–7826, June 2002.

[76] K. Golenberg, B. Kimelfeld, and Y. Sagiv. Keyword proximity search in complex data graphs. In *the Proceedings of the 28th ACM SIGMOD international conference on Management of data*, pages 927–940, New York, NY, USA, 2008. ACM.

[77] J. Graupmann. *The SphereSearch Engine for Graph-based Search on heterogeneous semi-structured data*. PhD thesis, Saarland University, 2006.

[78] J. Graupmann, R. Schenkel, and G. Weikum. The SphereSearch engine for unified ranked retrieval of heterogeneous XML and web documents. In *the Proceedings of the 31st international conference on Very large data bases (VLDB)*, pages 529–540. VLDB Endowment, 2005.

[79] B. K. Gregory, G. Marton, G. Borchardt, A. Brownell, S. Felshin, D. Loreto, J. Louis-rosenberg, B. Lu, F. Mora, S. Stiller, Ö. Uzuner, and A. Wilcox. External knowledge sources for question answering. In *the Proceedings of the 14th Annual Text Retrieval Conference (TREC)*. NIST, 2005.

[80] J. Han, X. Yan, and P. Yu. Mining and searching graphs and structures, Tutorial. In *12th ACM Conference on Knowledge Discovery and Data Mining (KDD)*, New York, NY, USA, 2006. ACM.

[81] E. Hartuv and R. Shamir. A clustering algorithm based on graph connectivity. *Information Processing Letters*, 76(4-6):175–181, 2000.

[82] H. He, H. Wang, J. Yang, and P. S. Yu. BLINKS: ranked keyword searches on graphs. In *the Proceedings of the 2007 ACM SIGMOD international conference on Management of data*, pages 305–316, New York, NY, USA, 2007. ACM.

[83] D. Hiemstra and A. P. de Vries. Relating the new language models of information retrieval to the traditional retrieval models. Technical report, Centre for Telematics and Information Technology, University of Twente, Enschede, Netherlands, 2000.

Bibliography

[84] I. Horrocks and P. F. Patel-Schneider. Reducing owl entailment to description logic satisfiability. *Journal of Web Semantics*, 1(4):17–29, 2003.

[85] V. Hristidis, L. Gravano, and Y. Papakonstantinou. Efficient ir-style keyword search over relational databases. In *the Proceedings of the 29th international conference on Very large data bases (VLDB)*, pages 850–861. VLDB Endowment, 2003.

[86] V. Hristidis and Y. Papakonstantinou. Discover: keyword search in relational databases. In *the Proceedings of the 28th international conference on Very Large Data Bases (VLDB)*, pages 670–681. VLDB Endowment, 2002.

[87] E. Ihler. Bounds on the quality of approximate solutions to the group steiner problem. In *the Proceedings of the 16th international workshop on Graph-theoretic concepts in computer science (WG)*, pages 109–118, New York, NY, USA, 1991. Springer-Verlag New York, Inc.

[88] I. F. Ilyas, G. Beskales, and M. A. Soliman. A survey of top-k query processing techniques in relational database systems. *ACM Computing Surveys*, 40(4):1–58, 2008.

[89] P. G. Ipeirotis, E. Agichtein, P. Jain, and L. Gravano. Towards a query optimizer for text-centric tasks. *ACM Transaction Database Systems*, 32(4):21, 2007.

[90] K. Järvelin and J. Kekäläinen. IR evaluation methods for retrieving highly relevant documents. In *the Proceedings of the 23rd annual international ACM SIGIR conference on Research and development in information retrieval (SIGIR)*, pages 41–48, New York, NY, USA, 2000. ACM.

[91] T. S. Jayram, R. Krishnamurthy, S. Raghavan, S. Vaithyanathan, and H. Zhu. AVATAR information extraction system. *IEEE Data Engineering Bulletin*, 29(1):40–48, 2006.

[92] V. Kacholia, S. Pandit, S. Chakrabarti, S. Sudarshan, R. Desai, and H. Karambelkar. Bidirectional expansion for keyword search on graph databases. In *the Proceedings of the 31st international conference on Very large data bases (VLDB)*, pages 505–516. VLDB Endowment, 2005.

[93] G. Kasneci, S. Elabssuoni, and G. Weikum. MING: mining informative entity-relationship subgraphs. In *the Proceedings of the 18th ACM Conference on Information and Knowledge Management (CIKM)*, New York, NY, USA, 2009. ACM.

[94] G. Kasneci, S. Elbassuoni, and G. Weikum. MING: mining informative entity-relationship subgraphs. Technical report, Max-Planck Institute for Informatics, Saarbruecken, Germany, 2009.

[95] G. Kasneci, M. Ramanath, M. Sozio, F. M. Suchanek, and G. Weikum. STAR: Steiner-tree approximation in relationship graphs. In *the Proceedings of the 2009 IEEE International Conference on Data Engineering (ICDE)*, pages 868–879, Washington, DC, USA, 2009. IEEE Computer Society.

[96] G. Kasneci, M. Ramanath, F. Suchanek, and G. Weikum. The yago-naga approach to knowledge discovery. *SIGMOD Record*, 37(4):41–47, 2008.

[97] G. Kasneci, F. M. Suchanek, G. Ifrim, S. Elbassuoni, M. Ramanath, and G. Weikum. NAGA: harvesting, searching and ranking knowledge. In *the Proceedings of the 28th ACM SIGMOD international conference on Management of data*, pages 1285–1288, New York, NY, USA, 2008. ACM.

[98] G. Kasneci, F. M. Suchanek, G. Ifrim, M. Ramanath, and G. Weikum. Naga: Searching and ranking knowledge. In *the Proceedings of the 24th International Conference on Data Engineering (ICDE)*, pages 953–962, Cancun, Mexico, 2008. IEEE Computer Society.

[99] G. Kasneci, F. M. Suchanek, M. Ramanath, and G. Weikum. How NAGA uncoils: searching with entities and relations. In *the Proceedings of the 16th international conference on World Wide Web (WWW)*, pages 1167–1168, New York, NY, USA, 2007. ACM.

[100] D. Kempe, J. Kleinberg, and É. Tardos. Maximizing the spread of influence through a social network. In *the Proceedings of the 9th ACM SIGKDD international conference on Knowledge discovery and data mining (KDD)*, pages 137–146, New York, NY, USA, 2003. ACM.

[101] B. Kimelfeld and Y. Sagiv. Finding and approximating top-k answers in keyword proximity search. In *the Proceedings of the 25th ACM SIGMOD-SIGACT-SIGART symposium on Principles of database systems (PODS)*, pages 173–182, New York, NY, USA, 2006. ACM.

[102] J. M. Kleinberg. Hubs, authorities, and communities. *ACM Computing Surveys*, 31(4):5–8.

[103] J. M. Kleinberg. Authoritative sources in a hyperlinked environment. *Journal of the ACM*, 46(5):604–632, 1999.

[104] L. Kou, G. Markowsky, and L. Berman. A fast algorithm for steiner trees. *Acta Informatica (Historical Archive)*, 15 (2), June 1981.

[105] M. Krötzsch, D. Vrandecic, M. Völkel, H. Haller, and R. Studer. Semantic wikipedia. *Journal of Web Semantics*, 5(4):251–261, 2007.

[106] R. Kumar, P. Raghavan, S. Rajagopalan, and A. Tomkins. Trawling the web for emerging cyber-communities. *Computer Networks*, 31(11-16):1481–1493, 1999.

[107] M. Kuramochi and G. Karypis. Frequent subgraph discovery. In *the Proceedings of the 2001 IEEE International Conference on Data Mining (ICDM)*, pages 313–320, Washington, DC, USA, 2001. IEEE Computer Society.

[108] J. Lafferty and C. Zhai. Document language models, query models, and risk minimization for information retrieval. In *the Proceedings of the 24th annual international ACM SIGIR conference on Research and development in information retrieval (SIGIR)*, pages 111–119, New York, NY, USA, 2001. ACM.

[109] J. Lehmann, J. Schüppel, and S. Auer. Discovering unknown connections – the DBpedia relationship finder. In *the Proceedings of 1st Conference on Social Semantic Web (CSSW)*, LNI, pages 99–110. GI, 2007.

[110] U. Leser. A query language for biological networks. *Bioinformatics*, 21(2):33–39, 2005.

[111] G. Li, B. C. Ooi, J. Feng, J. Wang, and L. Zhou. EASE: an effective 3-in-1 keyword search method for unstructured, semi-structured and structured data. In *the Proceedings of the 28th ACM SIGMOD international conference on Management of data*, pages 903–914, New York, NY, USA, 2008. ACM.

[112] W.-S. Li, K. S. Candan, Q. Vu, and D. Agrawal. Retrieving and organizing web pages by "information unit". In *the Proceedings of the 10th international conference on World Wide Web (WWW)*, pages 230–244, New York, NY, USA, 2001. ACM.

[113] W.-S. Li, K. S. Candan, Q. Vu, and D. Agrawal. Query relaxation by structure and semantics for retrieval of logical web documents. *IEEE Transactions on Knowledge and Data Engineering*, 14(4):768–791, 2002.

[114] D. Lin and P. Pantel. DIRT: discovery of inference rules from text. In *the Proceedings of the 7th ACM SIGKDD international conference on Knowledge discovery and data mining (KDD)*, pages 323–328, New York, NY, USA, 2001. ACM.

[115] X. Liu and B. W. Croft. Statistical language modeling for information retrieval. *Annual Review of Information Science and Technology*, 39(1):1–31, 2005.

Bibliography

[116] J. Madhavan, S. Cohen, X. Dong, A. Halevy, S. Jeffery, D. Ko, and C. Yu. Navigating the seas of structured web data. In *the Proceedings of the 3rd Biennial Conference on Innovative Data Systems Research (CIDR)*. www.crdrdb.org, 2007.

[117] M. E. Maron and J. L. Kuhns. On relevance, probabilistic indexing and information retrieval. *Journal of the ACM*, 7(3):216–244, 1960.

[118] K. Mehlhorn. A faster approximation algorithm for the steiner problem in graphs. *Information Processing Letters*, 27(3), 1988.

[119] T. Neumann and G. Weikum. RDF-3X: a RISC-style engine for RDF. *the Proceedings of the VLDB Endowment*, 1(1), 2008.

[120] Z. Nie, Y. Ma, S. Shi, J.-R. Wen, and W.-Y. Ma. Web object retrieval. In *the Proceedings of the 16th international conference on World Wide Web (WWW)*, pages 81–90, New York, NY, USA, 2007. ACM.

[121] I. Niles and A. Pease. Towards a standard upper ontology. In *the Proceedings of the international conference on Formal Ontology in Information Systems (FOIS)*, pages 2–9, New York, NY, USA, 2001. ACM.

[122] openRDF.org. Home of sesame. http://www.openrdf.org/index.jsp. Accessed 01-June-2009.

[123] A. N. Papadopoulos, A. Lyritsis, and Y. Manolopoulos. Skygraph: an algorithm for important subgraph discovery in relational graphs. *Data Mining and Knowledge Discovery*, 17(1):57–76, 2008.

[124] D. Petkova and W. B. Croft. Hierarchical language models for expert finding in enterprise corpora. In *the Proceedings of the 18th IEEE International Conference on Tools with Artificial Intelligence (ICTAI)*, pages 599–608, Washington, DC, USA, 2006. IEEE Computer Society.

[125] C. Plake, T. Schiemann, M. Pankalla, J. Hakenberg, and U. Leser. Ali baba: PubMed as a graph. *Bioinformatics*, 22(19), 2006.

[126] J. M. Ponte and W. B. Croft. A language modeling approach to information retrieval. In *the Proceedings of the 21st annual international ACM SIGIR conference on Research and development in information retrieval (SIGIR)*, pages 275–281, New York, NY, USA, 1998. ACM.

[127] S. P. Ponzetto and M. Strube. Deriving a large-scale taxonomy from wikipedia. In *the Proceedings of 22nd International Conference on Artificial Intelligence (AAAI)*, pages 1440–1445, Vancouver, British Columbia, Canada, 2007. AAAI Press.

Bibliography

[128] C. Ramakrishnan, W. H. Milnor, M. Perry, and A. P. Sheth. Discovering informative connection subgraphs in multi-relational graphs. *SIGKDD Explorations Newsletter*, 7(2):56–63, 2005.

[129] G. Reich and P. Widmayer. Beyond steiner's problem: a vlsi oriented generalization. In *the Proceedings of the 15th international workshop on Graph-theoretic concepts in computer science (WG)*, pages 196–210, New York, NY, USA, 1990. Springer-Verlag New York, Inc.

[130] S. Sarawagi. Information extraction. *Foundations and Trends in Databases*, 1(3):261–377, 2008.

[131] M. Sayyadian, H. LeKhac, A. Doan, and L. Gravano. Efficient keyword search across heterogeneous relational databases. In *the Proceedings of the 23rd International Conference on Data Engineering (ICDE)*, pages 346–355, Los Alamitos, USA, 2007. IEEE Computer Society.

[132] R. Schenkel, A. Theobald, and G. Weikum. Semantic similarity search on semistructured data with the XXL search engine. *Information Retrieval*, 8(4):521–545, 2005.

[133] P. Serdyukov and D. Hiemstra. Modeling documents as mixtures of persons for expert finding. In *the Proceedings of the 30th European Conference on IR Research (ECIR)*, Lecture Notes in Computer Science, pages 309–320. Springer Verlag, 2008.

[134] W. Shen, A. Doan, J. F. Naughton, and R. Ramakrishnan. Declarative information extraction using datalog with embedded extraction predicates. In *the Proceedings of the 33rd international conference on Very large data bases (VLDB)*, pages 1033–1044. VLDB Endowment, 2007.

[135] Stern.de. Test: wikipedia schlaegt brockhaus. `http://www.stern.de/computer-technik/internet/:stern-Test-Wikipedia-Brockhaus/604423.html`. Accessed 01-June-2009.

[136] F. Suchanek, G. Kasneci, and G. Weikum. YAGO - a large ontology from wikipedia and wordnet. *Journal of Web Semantics*, 6(3):203–217, 2008.

[137] F. M. Suchanek. *Automated Construction and Growth of a Large Ontology*. PhD thesis, Saarland University, 2008.

[138] F. M. Suchanek, G. Kasneci, and G. Weikum. YAGO: a core of semantic knowledge. In *the Proceedings of the 16th international conference on World Wide Web (WWW)*, pages 697–706, New York, NY, USA, 2007. ACM.

Bibliography

[139] H. Tong and C. Faloutsos. Center-piece subgraphs: problem definition and fast solutions. In *the Proceedings of the 12th ACM SIGKDD international conference on Knowledge discovery and data mining (KDD)*, pages 404–413, New York, NY, USA, 2006. ACM.

[140] H. Tong, C. Faloutsos, and J.-Y. Pan. Fast random walk with restart and its applications. In *the Proceedings of the 6th International Conference on Data Mining (ICDM)*, pages 613–622, Washington, DC, USA, 2006. IEEE Computer Society.

[141] S. Trißl and U. Leser. Fast and practical indexing and querying of very large graphs. In *the Proceedings of the 27th ACM SIGMOD international conference on Management of data*, pages 845–856, New York, NY, USA, 2007. ACM.

[142] D. Vallet and H. Zaragoza. Inferring the most important types of a query: a semantic approach. In *the Proceedings of the 31st annual international ACM SIGIR conference on Research and development in information retrieval (SIGIR)*, pages 857–858, New York, NY, USA, 2008. ACM.

[143] W. Weerkamp, K. Balog, and E. J. Meij. A generative language modeling approach for ranking entities. In *Advances in Focused Retrieval*, Lecture Notes in Computer Science, Berlin / Heidelberg, 2009. Springer.

[144] G. Weikum. Information retrieval and data mining. Computer Science Lecture at University of Saarland, Winter Term, 2007-2008.

[145] G. Weikum, G. Kasneci, M. Ramanath, and F. Suchanek. Database and information-retrieval methods for knowledge discovery. *Communications of the ACM (CACM)*, 52(4):56–64, 2009.

[146] D. S. Weld, R. Hoffmann, and F. Wu. Using wikipedia to bootstrap open information extraction. *SIGMOD Record*, 37(4), 2008.

[147] K. Wilkinson, C. Sayers, H. A. Kuno, and D. Reynolds. Efficient RDF storage and retrieval in Jena2. In *the Proceedings of the 1st International Workshop on Semantic Web and Databases (SWDB)*, pages 35–43, 2003.

[148] F. Wu and D. S. Weld. Autonomously semantifying wikipedia. In *the Proceedings of the sixteenth ACM conference on Conference on information and knowledge management (CIKM)*, pages 41–50, New York, NY, USA, 2007. ACM.

[149] F. Wu and D. S. Weld. Automatically refining the wikipedia infobox ontology. In *Proceeding of the 17th international conference on World Wide Web (WWW)*, pages 635–644, New York, NY, USA, 2008. ACM.

Bibliography

[150] Yahoo. Yahoo! answers. http://answers.yahoo.com/, 2005. Accessed 01-June-2009.

[151] X. Yan, X. J. Zhou, and J. Han. Mining closed relational graphs with connectivity constraints. In *the Proceedings of the eleventh ACM SIGKDD international conference on Knowledge discovery in data mining (KDD)*, pages 324–333, New York, NY, USA, 2005. ACM.

[152] C. Zhai and J. Lafferty. A risk minimization framework for information retrieval. *Information Processing and Management*, 42(1):31–55, 2006.

[153] J. Zhu, Z. Nie, J.-R. Wen, B. Zhang, and W.-Y. Ma. Simultaneous record detection and attribute labeling in web data extraction. In *the Proceedings of the 12th ACM SIGKDD international conference on Knowledge discovery and data mining (KDD)*, pages 494–503, New York, NY, USA, 2006. ACM.

Die VDM Verlagsservicegesellschaft sucht für wissenschaftliche Verlage abgeschlossene und herausragende

Dissertationen, Habilitationen, Diplomarbeiten, Master Theses, Magisterarbeiten usw.

für die kostenlose Publikation als Fachbuch.

Sie verfügen über eine Arbeit, die hohen inhaltlichen und formalen Ansprüchen genügt, und haben Interesse an einer honorarvergüteten Publikation?

Dann senden Sie bitte erste Informationen über sich und Ihre Arbeit per Email an *info@vdm-vsg.de*.

Sie erhalten kurzfristig unser Feedback!

VDM Verlagsservicegesellschaft mbH
Dudweiler Landstr. 99　　　　　Telefon　+49 681 3720 174
D - 66123 Saarbrücken　　　　　Fax　　　+49 681 3720 1749
www.vdm-vsg.de

Die VDM Verlagsservicegesellschaft mbH vertritt

Printed by Books on Demand GmbH, Norderstedt / Germany